TECH SEO GUIDE

A REFERENCE GUIDE FOR DEVELOPERS AND MARKETERS INVOLVED IN TECHNICAL SEO

Matthew Edgar

Tech SEO Guide: A Reference Guide for Developers and Marketers Involved in Technical SEO

Matthew Edgar
Centennial, CO, USA

ISBN-13 (pbk): 978-1-4842-9053-8   ISBN-13 (electronic): 978-1-4842-9054-5
https://doi.org/10.1007/978-1-4842-9054-5

Managing Director, Apress Media LLC: Welmoed Spahr
Acquisitions Editor: Shiva Ramachandran
Development Editor: James Markham
Coordinating Editor: Jessica Vakili

Distributed to the book trade worldwide by Springer Science+Business Media New York, 1 New York Plaza, New York, NY 10004. Phone 1-800-SPRINGER, fax (201) 348-4505, e-mail orders-ny@springer-sbm.com, or visit www.springeronline.com. Apress Media, LLC is a California LLC and the sole member (owner) is Springer Science + Business Media Finance Inc (SSBM Finance Inc). SSBM Finance Inc is a **Delaware** corporation.

For information on translations, please e-mail booktranslations@springernature.com; for reprint, paperback, or audio rights, please e-mail bookpermissions@springernature.com.

Apress titles may be purchased in bulk for academic, corporate, or promotional use. eBook versions and licenses are also available for most titles. For more information, reference our Print and eBook Bulk Sales web page at http://www.apress.com/bulk-sales.

Any source code or other supplementary material referenced by the author in this book is available to readers on the Github repository: https://github.com/Apress/Tech-SEO-Guide. For more detailed information, please visit http://www.apress.com/source-code.

Printed on acid-free paper

Contents

About the Author

Matthew Edgar is a consultant and partner at Colorado-based Elementive (www.elementive.com), a consulting firm specializing in technical SEO. Matthew has worked in the web performance field since 2001 and holds a master's degree in information and communications technology from the University of Denver. He has been interviewed by Forbes, American Express, and other publications about web analytics, CRO, UX, and SEO. He regularly speaks at conferences and teaches workshops, including speaking at MozCon, SMX, and MarTech and teaching courses with O'Reilly Media. Learn more about or connect with Matthew at MatthewEdgar.net.

About the Technical Reviewer

April Ursula Fox comes from a background in business, general and social media marketing, data analytics, and educational psychology, delivering analytics solutions and specialized training to global organizations and marketing agencies around the world. With names such as AutoCAD, Oracle, Samsung, Peugeot & Citroen, Danone, Nestle, Ogilvy, Leo Burnett, and Havas, among many others, April has expanded the understanding of data gathering and analysis with significant impact to the workflow and performance of data-infused teams.

Today, April is moving further into the educational aspect of her work, dedicated to studies in educational psychology and all of the components that affect human learning and growth. Within this new focus, April understands that despite our incredible technological evolution, there is still a key human component that will ultimately make or break the performance of an individual or a team. Such interest moves April into the pursuit of the understanding of the intersectionality of the components of the human condition, or how all aspects of our lives will affect us within different contexts of performance. These studies will likely inform our future technological developments and the way we learn and become experts in using technology to enhance our living experience.

Introduction

What Is Technical (Tech) SEO?

Technical SEO, or tech SEO, is the ongoing process of optimizing a website's code and server configuration to better communicate with search engine robots. The goal is to make it easier for robots to crawl the website so that robots can more easily index the page contained on the website. The more easily robots can crawl and index a website's pages, the more likely it is that the website will earn higher rankings in organic search results and drive more visitors to the website from those results.

Beyond SEO…

The process of technical SEO, however, requires thinking about more than search engine robots. Optimizing a website's technical structure also improves the experience humans have when visiting the website. Websites that load fast will be easier for robots to crawl and will likely receive a boost in rankings, but faster loading websites also create a better experience for visitors, resulting in greater engagement and conversions. The same is true for websites that contain fewer errors—fewer errors mean fewer problems for both robots and human visitors. Websites that offer more robust navigation paths will help robots find and index more pages on the website but will also help visitors find and engage with more pages. Removing duplicate or low-quality content will lead to better rankings but will also reduce the chance visitors will be driven away from the website by that content. Making a website's design friendlier on mobile devices will help comply with Google's mobile ranking factors and will also help visitors engage with the website on mobile devices.

In short, improving the technical factors discussed has tremendous benefits to the website's overall performance beyond higher rankings and traffic. While this book focuses on the search-related benefits of improving the website's technical structure, it is important to remember there are many other benefits to consider and measure as well.

Why This Book?

This book is meant to serve as a reference guide for the wide variety of people who work on tech SEO factors: SEO professionals, marketers working outside of SEO, developers, server engineers, designers, and more. This book can be used by an experienced SEO needing to quickly jog their memory on a complex topic or by somebody new to tech SEO needing help with the basics. Whatever your current level of tech SEO experience, my hope is this book is a helpful resource as you optimize your website and organic search presence.

Examples   There are several spreadsheets and code samples shared throughout this guide. To make it easier to work with these spreadsheets and code samples, an XLS template of each table and code examples can be accessed at ___.

Crawling and Indexing

This chapter covers crawling and indexing. This is where SEO begins. Search engine robots need to crawl a website's pages to find all the information contained on those pages. Once crawled, robots then process and store the information from the website in the search engine's index. If robots are unable to crawl or index a website's pages, the website may not rank in search results. Achieving optimal SEO performance requires understanding how search engine robots work and ensuring that robots are able to crawl and index a website effectively.

What Are Search Robots?

The term "robots" is a convenient way to refer to a complex collection of different programs that search engines use to understand and evaluate websites. It is helpful to group the multitude of programs into four main groups. There are programs designed to

- Crawl content
- Fetch content
- Analyze and assess the content
- Rank that content in search results

© Matthew Edgar 2023
M. Edgar, *Tech SEO Guide*, https://doi.org/10.1007/978-1-4842-9054-5_1

Programs designed to crawl are primarily focused on finding every file contained on every website, continually looking for new or updated files. Search engine crawlers want to find every type of file, including HTML pages, PDFs, images, videos, and more. Crawling programs find files through links, including links on the website (internal links) and links on other websites (external links). The more links, internal or external, that reference a page, the more likely it is that robots will find and crawl that page.

After finding the files, different programs fetch the content contained within those files and pass the content along to other programs for further analysis. The cleaner a website's code and the better optimized the website's server configuration, the easier it is for these programs to fetch the content from the crawled file. The fetching programs also extract links contained in the fetched file and feed those links back into the crawling programs (the importance of links will be discussed more throughout this book, in particular in Chapter 3).

The next set of programs analyze and assess the content and, following this assessment and analysis, index the content for easier retrieval. These assessment and analysis programs are designed to extract meaning from content. This is the realm of natural language processing and machine learning. This is also where search engines determine the quality of the content and, by extension, the quality of the website overall. Assessment and analysis programs also consider external factors, like links or citations from external websites, to establish a website's authority.

Finally, there are programs designed to review the extracted information contained in the index to determine where the file ought to rank in search results (or if it ought to rank at all). If a page from a website is selected to rank in search results, this group of programs evaluates any schema markup found on the page to determine if search result listings should be enhanced in some way—such as adding stars for a review (see Chapter 4 for more about schema). These programs also consider factors like a page's speed to determine if a page should be boosted a bit higher in rankings than another website (see Chapter 6 for more about speed).

Many of these activities will be discussed in greater detail in the chapters ahead. When the term "robots" is used throughout this book, it is a shorthand reference to all these complex activities.

Search Robot Operations

Given the complexity of search robots, it can be easier to break the vast number of programs into the two main operations robots take to understand and evaluate a website:

- **Crawling**: The robot's first step is discovering that a website exists and fetching the content from that website. Once the robot discovers the website, it crawls through every file on the website it can find. Websites are discovered and files are found predominately via links. Limits can be placed on which files a robot should be able to access. The primary SEO crawling goal is to ensure a robot can successfully find everything it should while finding nothing it shouldn't.

- **Indexing**: During the crawl, robots add all the files found to a database, but after the crawling is complete, robots decide how to index, or organize, all the files found. The first step is to exclude certain types of files, like error pages or pages robots have been specifically told not to index. Next, robots extract and evaluate the information from the file. Based on the information found, robots determine if the indexed page should or should not be allowed to rank within search results. The primary SEO indexing goal is to ensure a robot is properly evaluating the files found on the website, especially ensuring that exclusions are properly handled.

What Is Crawl Budget Optimization?

Crawl budget represents two different numbers: the number of files that a search engine robot has crawled on a website and the number of files that a search engine robot should crawl on a website. To begin understanding a website's crawl budget, both numbers need to be measured:

- **Files that have been crawled**: It is easiest to count how many files a robot has crawled within the website's log files. However, this can also be gauged by evaluating how many pages are currently indexed in search results.

- **Files that should be crawled**: This number of files that robots could crawl is not necessarily the same as the number of total files contained on a website. Rather, this number reflects the number of files on a website that it is important for a search engine to crawl. There will be pages on most websites that robots should not crawl, such as on-site search pages or pages related to a website's shopping cart.

Once measured, the numbers can be compared. If there are 1000 files on a website that robots should crawl, but robots have only crawled 500 of those files, then work needs to be done to determine why the other files could not be crawled. Similarly, if only 1000 files on a website should be crawled, but robots crawled 2000 files, then work needs to be done to determine what the extra files are and why robots crawled those files.

What Is Mobile-First Indexing?

Mobile-first indexing means Googlebot crawls and evaluates the mobile website first and uses what is found on the mobile website to decide how to index the website's pages and where to rank those pages in search results. This is Google's default method of crawling and indexing a website. Googlebot will still crawl the desktop website but usually not as often as the mobile website. What robots find on the desktop website may not have as much influence on what is ranked in search results.

The most important consideration for mobile-first indexing is the equivalency between the desktop and mobile websites. If the robot only evaluated the mobile website, is there information only contained on the desktop website that would be missed? Do the mobile and desktop websites share the same content and serve the same general purpose? Some hidden content on mobile is expected, especially for decorative or supportive content, but the main content should communicate essentially the same information regardless of the device.

Methods to Guide Robots

The first common method of guiding a robot's activity on a website is the robots.txt file. This is a text file located in the website's root directory. For example, Google's robots.txt file is located at www.google.com/robots.txt.

The second common method of guiding a robot through a website is with a <meta> tag located in the head of a page with a name attribute of "robots" and the "content" attribute specifying what the robot can and cannot do on the page. The shorthand name for this tag is "meta robots."

An alternative but less common method is the X-Robots tag.

All of these methods are discussed in more detail through the remainder of this chapter.

Disallow: Robots.txt

The disallow directive is specified within the robots.txt file. With this directive, the website is stating that robots are not allowed to crawl this file or directory. Examples of a directory and file disallow:

```
Disallow: /a-secret-directory
Disallow: /a-secret-file.pdf
```

When specified, a search robot that respects and follows this directive will not crawl the disallowed file or directory. However, the disallowed files may still be indexed and appear in search results. If that seems odd, think back to the two main operations: crawling and indexing. A disallow directive only affects crawling but provides no direction to robots regarding what to do regarding how to index a disallowed file. One common example is that robots find links that reference the disallowed file, either on the website itself or on other websites. Given the volume of the links referencing the disallowed file, robots decide that the disallowed file must still be important, and, as a result, robots may place that file in the index and rank it in search results even if the file itself cannot be crawled.

Noindex: Meta Robots

The "noindex" directive can be specified on a page within the meta robots tag.

Example:

```
<meta name="robots" content="noindex" />
```

When specified on the robots.txt file, "noindex" does not prevent a robot from crawling the page, only from indexing it (assuming the search robot will follow the directive).

Note that the meta robots tag can be narrowed down to a specific robot. To only direct Googlebot not to index a page, the name attribute changes from "robots" to "googlebot." Other robots would be permitted to index this page. Example:

```
<meta name="googlebot" content="noindex" />
```

Using Noindex and Disallow Directives

It is important to be clear on how the disallow and noindex directives work together. There are three ways these directives can be combined to affect indexing and crawling as shown in Table 1-1.

Table 1-1. Noindex and Disallow Scenarios

	Robots.txt Disallow	Meta Robots Noindex
Scenario 1		X
Scenario 2	X	
Scenario 3	X	X

In Scenario 1, the page with a noindex setting will be excluded from the index and, as a result, not included in a search result. However, a robot may still crawl this page, meaning the robots can access content on the page and follow links on the page.

In Scenario 2, the page will not be crawled but may be indexed and appear in search results. Because the robot did not crawl the page, the robot knows nothing about the page's content. Any information robots have gathered about this page in search result will be gathered from other sources, like external or internal links to the disallowed page.

Scenario 3 will operate exactly like Scenario 2. This is because when a disallow is specified, a robot will not crawl the page. If the robot does not crawl to the page, the robot will not fetch or process any code contained on this page, which includes the meta tag directing robots not to index the page. If a page needs to be set to noindex and disallowed, set the noindex first, then after the page is removed from the search index, set the disallow (also see section "Forbidding Access" later in this chapter).

Index or Noindex?

Generally, it should be left up to robots to decide what should or should not be indexed. However, some pages may be good candidates to noindex, such as the following:

- **Low-quality pages**: These pages may not make for good entry points from a search result as they would not satisfy a human visitor's interests and intentions.

- **Category or tag pages on a blog**: It might be better for people to find a relevant blog post from the search result instead of category or tag pages.

- **Landing pages**: A landing page for an email or advertisement campaign should not be indexed or ranked in organic search results.

- **Duplicate content**: If two pages share the same content, one of the pages can be set to noindex to prevent the pages from competing against each other to appear in the same search results. See Chapter 3 for more about duplicate content.

Disallow or Allow?

Disallow directives should be used sparingly. It is easy to disallow more than intended. However, on larger websites, the disallow directive can help limit how much robots crawl. This keeps the robot focused on pages that matter as well as helping to manage limited server capacity.

For example, if the website includes an internal search engine, the pages associated with that internal search engine could be blocked from crawling via a robots.txt disallow directive since it would not be particularly helpful for a robot to see these internal search pages.

Forbidding Access

A robot can ignore disallow or noindex directives specified in the robots.txt or meta robots tag. Along with search robots, malicious robots crawl websites, and those robots will certainly not obey requests to ignore certain files or links.

Certain files, directories, or pages must be hidden from view with no exceptions—such as staging websites, login areas, or files that contain restricted information. In these cases, blocking access via a disallow or noindex directive is not the right solution.

The better option is to require a login to access the page or only permitting access to users with specific IP addresses. If a visitor cannot log in or uses an unauthorized IP address, the requested file should return a status code of 403 Forbidden.

Malicious robots can be blocked with firewall rules. When setting these rules, ensure that Googlebot and Bingbot are not blocked.

Handling Staging Environments

Staging and development environments should always be restricted from view via password protection or other forms of authentication. These types of environments are often very similar to the main website. As a result, if robots find these environments, they can mistake the staging or development website as a duplicate copy of the main website. If this happens, this can cause the main website to lose rankings in search results. Even setting aside those problems, the staging and development environments are only intended for very specific audiences and should not be shared with search robots.

Do Not Block JavaScript (JS), Cascading Style Sheets (CSS), Images, or Videos

To understand a page, robots need to fetch all parts of the page, including CSS, JavaScript, images, and videos. CSS files alter the website's layout for mobile and desktop devices. JavaScript can load or alter a page's content. Images and videos are often a critical piece of a page's content. Images and videos can also rank in search results and drive visitors to the website. As a result, these files should not be blocked from crawling and, in the case of images and videos, should not be blocked from indexing.

Nofollow: Meta Robots

Along with providing direction on indexing, the content attribute of the meta robots tag can specify whether any links contained on a page should be followed (crawled) by a search robot or if the links on the page should not be followed (crawled). There are two options for the meta robots follow directive:

- **follow**: Tells a robot they can crawl any links contained on the page
- **nofollow**: Tells a robot they cannot crawl links contained on a page

Usually, nofollow or follow is specified along with a noindex. Examples:

```
<meta name="robots" content="noindex,follow" />
<meta name="robots" content="noindex,nofollow" />
```

The meta robots nofollow only instructs robots not to follow links contained on a specific page. If another page contains the same links but that page does not have the nofollow directive, those links will be followed.

Follow or Nofollow and Page Sculpting

Generally, robots should be told they can follow all links on a page. Being too aggressive in specifying which links to follow or nofollow via the meta robots tag can begin to look as if the website is attempting to manipulate a robot's perception of a website. This is a practice known as page sculpting, where nofollow directives are used to manipulate which signals are passed between website pages. At best, these attempts to manipulate a robot with page sculpting no longer work. At worst, attempts to manipulate robots in this way can lead to a penalty or manual action, removing a website from search results.

Link Qualifiers and Rel Nofollow

A nofollow can also be specified at a link level using the rel attribute inside an <a> element. This type of nofollow does not influence whether the link is crawled. Instead, this nofollow explains (or qualifies) the nature of why a given link is included on a page. These qualifiers only need to be used on external links (links pointing to a different website).

Traditionally, rel="nofollow" was used to specify any links that were sponsored or had a monetary relationship. Example of rel="nofollow":

```
<a href="/no-robots-here" rel="nofollow">Link</a>
```

Along with nofollow, Google supports two other values that can qualify the nature of the link: sponsored and ugc. The rel="sponsored" qualifier is for any paid link, and rel="ugc" indicates the link is part of user-generated content. The nofollow qualifier can still be used either alongside or instead of the sponsored and ugc qualifiers. Example:

```
<a href="/no-robots-here" rel="ugc,nofollow">User Generated Link</a>
<a href="/no-robots-here" rel="sponsored,nofollow">Paid Link</a>
```

X-Robots-Tag

Directives to guide robots can also be specified in the HTTP header using the X-Robots-Tag. This operates similar to the meta robots tag, allowing for control over indexing with the noindex statement and control over crawling links contained on that page via nofollow. Like the meta robots tag and the robots. txt file, these directives are not mandatory for a robot to follow. Unlike the meta robots tag, the X-Robots HTTP header can be specified on non-HTML files, like PDFs or images. Examples:

```
X-Robots-Tag: noindex
X-Robots-Tag: noindex,follow
X-Robots-Tag: noindex,nofollow
```

In Practice: Troubleshooting Google Not Crawling or Indexing Pages

Why do some pages not rank in search results? This is one of the more common SEO problems companies face, and there is no easy answer. A lack of rankings can be due to many different issues or a combination of issues. A lot of an SEO team's time is spent investigating which issues are present and which issues are to blame.

Checking for indexing or crawling issues is a good place to begin this investigation as it represents one of the most fundamental SEO issues. If robots are unable to crawl a website's pages, then those pages will be excluded from the search engine's index. If the pages are crawled but cannot be properly analyzed or assessed by search engines, then those pages will be excluded from the search engine's index. Any pages excluded from the search engine's index cannot rank in search results.

Even if crawling or indexing problems are not present, reviewing the website for these types of issues can rule out crawling or indexing as the issue that is preventing rankings. By ruling out crawling or indexing as the issue, the investigation into the lack of rankings can focus elsewhere and identify other issues. The following steps review how to tell if crawling or indexing issues are present on a website and contributing to a ranking drop.

Step 1: What Pages Should Be Crawled and Indexed?

Diagnosing problems with crawling or indexing requires having a complete list of all the pages on the website that should, ideally, rank in search results. Pages, in this case, refer to any URL on the website that should appear in search results—including HTML pages, images, and videos.

The easiest way to obtain this list is from the website's XML sitemap file. If properly configured, the XML sitemap file will contain a list of all the pages and files that should rank in search results. See Chapter 5 for more details about properly configuring the XML sitemap so that it provides an accurate list of URLs. If the XML sitemap is properly configured, then obtaining the complete list of all pages that should rank is a matter of copying all of the URLs from the XML sitemap. If the XML sitemap file is not properly configured, see Chapter 3's "In Practice" section which discusses a method of obtaining a list of all the pages contained on the website.

Once the full list of pages that should rank has been obtained, create a new spreadsheet and add this list of URLs to the first column of that spreadsheet. In the second column of the spreadsheet, note which pages are currently not ranking in search results. This spreadsheet should resemble Table 1-2.

Table 1-2. Example Spreadsheet to
Diagnose Crawling and Indexing Issues, Step 1

Page URL	Ranking?
/	
/products	
/products/blue-widget	No
/about-us	
/support	No
/support/instructions	No
/customer-login	No
/blog	No
/blog/a-great-post/	No

Step 2: What Noindex or Disallow Directives Currently Exist?

The next step is finding any robots.txt, meta robots, or X-Robots directives for each of those pages. If the page should rank in search results, then there should be no noindex or disallow directive present for the page. However, noindex or disallow directives can accidentally be added to the page or on the robots.txt file during website updates. That is likely the case if pages were ranking before and have suddenly stopped ranking. As a result, it is helpful to check these directives to rule out these types of accidental additions.

The meta robots and X-Robots information can be obtained by crawling a full list of all the URLs in a crawling tool. For example, in Screaming Frog, this can be done in List mode by uploading a text file containing all of the URLs. After Screaming Frog completes the crawl, review the meta robots and the X-Robots information under the Directives tab. Add a second column to the spreadsheet and list the information about the noindex directive to the spreadsheet for each URL, as shown in Table 1-3.

Table 1-3. Example Spreadsheet to Diagnose Crawling and Indexing Issues, Step 2—Noindex

Page URL	Ranking?	Noindex (Meta Robots or X-Robots)
/		
/products		
/products/blue-widget	No	
/about-us		
/support	No	
/support/instructions	No	
/customer-login	No	noindex
/blog	No	
/blog/a-great-post/	No	

Some crawl tools will also include robots.txt disallow directives related to each page. If that information is not provided by the crawl tool, these directives will need to be manually added. For example, if the robots.txt file disallows the /blog directory, then update the spreadsheet to note that those URLs contain a disallow.

Add a third column to the spreadsheet listing any disallow directives related to each page, as shown in Table 1-4.

Table 1-4. Example Spreadsheet to Diagnose Crawling and Indexing Issues, Step 2—Disallow

Page URL	Ranking?	Noindex (Meta Robots or X-Robots)	Disallow (Robots.txt)
/			
/products			
/products/blue-widget	No		
/about-us			
/support	No		
/support/instructions	No		
/customer-login	No	noindex	
/blog	No		disallow
/blog/a-great-post/	No		disallow

If there are noindex directives noted in the spreadsheet, then that can explain why a page is not ranking in search results. For example, the /customer-login page in the example spreadsheet is likely not ranking because of the noindex directive.

Pages blocked by a disallow can rank in search results, but typically will rank lower because robots are unable to crawl the page to determine what content the page contains. In the example spreadsheet, it is possible the disallow is preventing the pages in the /blog directory from ranking.

Before continuing with the rest of the steps reviewing for crawling or indexing issues, remove any noindex or disallow directives. If the ranking problem remains after removing the directives, or if no noindex or disallow directives were present, continue the analysis with Step 3.

Step 3: Are Pages Being Crawled?

After ruling out noindex or disallow directives as the cause, the next step is understanding how the pages are being crawled currently. The only way to find this information is within the website's log files. To spot trends, it is helpful to take a wider view on log files, such as looking a month's worth of crawls at a time.

The log files can be downloaded from the hosting company and opened in Excel or a specialty log file review program, such as Loggly or http Logs Viewer. Here is an example log file hit for Googlebot accessing the URL / some-sample-page:

```
12.345.67.890  -  -  [15/Nov/2022:19:57:26  -0700]  "GET  /some-sample-page
HTTP/1.1" 200 24594 "-" "Mozilla/5.0 (compatible; Googlebot/2.1; +http://www.
google.com/bot.html)"
```

Log files will contain hits for every visit to the website. To understand crawl issues, it is only important to see hits from search engine robots. Therefore, hits should be filtered to a specific user agent related to a particular search engine robot, such as filtering to Googlebot or Bingbot. Once filtered, the hit can now be referred to as a crawl—a crawl is a hit from a search engine robot. Count the number of hits or crawls to each page listed in the spreadsheet created in Step 1. Most log file programs will provide a way to export a list of these counts.

Add the number of crawls for each page to the spreadsheet in a new column, as shown in Table 1-5. These numbers will describe crawl patterns, and those crawl patterns in turn will help identify any crawl-related issues that may exist.

Table 1-5. Example Spreadsheet to Diagnose Crawling and Indexing Issues, Step 3

Page URL	Ranking?	Noindex	Disallow	Crawl Count
/				76
/products				48
/products/blue-widget	No			5
/about-us				59
/support	No			0
/support/instructions	No			20
/customer-login	No	noindex		25
/blog	No		disallow	0
/blog/a-great-post/	No		disallow	0

Search engine robots should be regularly crawling most of the pages that rank in search results. Robots need to check the pages that rank for any updates or changes to content and confirm the page should still continue to rank in search results. If the log files represent a month's worth of crawl activity, then nearly every page listed in the spreadsheet should have more than zero crawls.

There can be some outlier pages with zero crawls, especially on larger websites. It could be the page was added to the website only within the last few days, and robots simply have not found that page yet. Or, it could be the page is crawled at a lower frequency, such as crawled every six weeks instead of every month. To confirm this, log files for previous months can be checked to determine if and when a page was crawled before. This is also why it is helpful to make log file checks part of regular monthly or quarterly reporting, to understand how crawl activity trends over time.

If there are more than a few pages with zero crawls, that does indicate a bigger crawl problem could exist, especially if the pages with the zero crawls have remained at that zero-crawl level for several months. If a page gets zero crawls, that typically means robots do not see the importance in crawling that page, or that could mean robots are trying to crawl the page but are having some difficulty in doing so. In the example spreadsheet, the /support page has zero crawls. Because it is a prominent page on the website that has existed for some time and is regularly updated, it should be crawled. As a result, this might mean this page has a crawl-related issue present.

Crawl-related problems do not only apply to pages with zero crawls. Some pages may get a few crawls but not enough crawls, given the relative importance of the page—search robots are not crawling the page enough to adequately evaluate the page and determine if it should rank. For each page listed in the spreadsheet, identify any pages that have a lower number of crawls. In the example spreadsheet, the pages listed have an average of about 26 crawls within a month. In this example, the /products/blue-widget page is well below the average with only 5 crawls within the month—that crawl level is suspiciously low because this spreadsheet only represents important pages that should rank in search results and, therefore, should be crawled more often.

If there are pages that should rank in search results that are not getting crawled at all or enough, then the tests in Step 4 will help clarify if robots are unable to crawl the page. Along with conducting the crawl tests in Step 4, it is worth keeping in mind that crawls can also be lower if a page is orphaned (see Chapter 3 for more details).

Alternatively, there might be plenty of crawls of each page on the website, including a higher number of crawls for the pages that are not currently ranking in search results. This usually indicates that robots are not struggling to crawl the pages contained on the website and that the issue instead exists with decisions the search engine is making after crawling the page (discussed in Step 5). However, if a page is getting a high volume of crawls and is not ranking in search results, it is still worth testing the page further, in Step 4, to confirm the robot is successfully crawling the page.

Step 4: Can Robots Crawl the Page?

A crawl in a log file merely indicates a robot attempted to access a particular URL. It does not indicate if that crawl was successful, resulting in the robot fetching all of the important content contained on that page. The next step is to test if robots can successfully crawl a page and fetch the page's content.

Not every page needs to be tested. Instead, find a few representative pages for each page type. For example, instead of testing each individual blog post, test two or three blog posts. Assuming the blog posts use a common template, then a crawl issue present on two or three posts would indicate that issue is present on other posts using that same template.

One tool to test if the pages are crawlable is Google Search Console. In Google Search Console, inspect the URL that currently has zero or few crawls with a live test. After conducting the live test, review the HTML code that has been returned and review the screenshot of the page. Does the HTML code contain the correct content for the page? Does the screenshot look like it returned the correct page? When crawl issues are present, the HTML or the screenshot might be blank or might be missing critical pieces of content. This could be due to a number of issues, including problems with how the page is rendered that prevents robots from seeing the content (see Chapter 3) or slower speeds that make it difficult for robots to load the page during a crawl (see Chapter 6).

Another helpful tool is a headless browser. Unlike a regular web browser, a headless browser operates without a graphical user interface (or GUI). Googlebot and Bingbot both use headless browsers to crawl the Web, and using a headless browser can provide a way of mimicking what these search engine robots are doing. A page can be crawled in a headless browser to see what HTML code is returned and check if the HTML code contains the correct content. Numerous resources are available online to configure a headless browser, though doing so requires some familiarity working with code or a directive line interface.

Unlike Google Search Console, a headless browser can be configured to crawl a page in different ways helping to identify crawl-related issues. For example, the suspicion might be that slower speeds are preventing robots from being able to successfully crawl a particular page. In this case, using a directive line headless browser, the virtual-time-budget option can be included to allow a longer delay when loading a page. If the HTML content is returned successfully when the time is delayed, then that confirms a speed issue is present and causing robots to not crawl the page successfully, which could cause the page not to rank.

In the spreadsheet, add a new column and note any issues found when conducting the different crawl tests on each of the pages. As well, note any potential causes the crawl tests indicate. In the example spreadsheet shown in Table 1-6, the / support page did not return any content in the HTML, and the screenshot returned of the page was blank. These issues were not helped by adjusting the timing. That indicates an issue with how this page is rendered, and those issues might be an underlying cause of why this page is not ranking in search results.

Table 1-6. Example Spreadsheet to Diagnose Crawling and Indexing Issues, Step 4

Page URL	Ranking?	...	Crawl Count	Crawl Tests and Issues
/			76	
/products			48	
/products/ blue-widget	No		5	No crawl issues detected
/about-us			59	
/support	No		0	Blank page and no HTML returned in the live test and headless browser—not changed with time adjustments
/support/ instructions	No		20	No crawl issues detected
/customer-login	No		25	
/blog	No		0	
/blog/a-great-post/	No		0	

Step 5: What Index Issues Are Present?

If the page is being crawled successfully but still not ranking in search results, that might indicate an indexing problem. An indexing problem means that after crawling and analyzing a page, the robot found some reason to exclude that page. That could be due to a variety of factors, including content quality (see Chapter 3), errors that were present on the page (see Chapter 7), or because the page redirected somewhere else (see Chapter 8). The next step then is to see if and why robots are excluding a page.

To begin, determine if the page is even indexed. The easiest way is to use a site: query. In Google or Bing, type "site:" followed by the page URL in the search box and then conduct the search. For example, searching "site: `https://www.matthewedgar.net/about-matthew-edgar/`" would indicate if my website's about page is included in the search engine's index. If the page is returned after checking the site: query, then there are no indexing issues present. If the page is not returned, then there is possibly an indexing issue.

Another tool to understand indexing behavior and related problems is Google Search Console. Under the Pages report (previously called Coverage), Google Search Console shows which pages are not indexed and provides reasons why. Remember, it is perfectly acceptable that some pages will not be included in the index. It only is a problem if the pages not in the index are pages that should be in the index.

Begin by reviewing each reason a page is Not Indexed listed in Google Search Console. Under each reason category, review which URLs are listed. Of the URLs listed, do any match a URL for a page that should rank in search results? If so, note the reason why the page is not indexed in the spreadsheet. In the example spreadsheet, the /products/blue-widget page did have an indexing issue reported in Google Search Console related to duplication. This is a common issue that can prevent pages from being indexed. Duplication issues are discussed in detail in Chapter 3.

Not all URLs will be listed in the Not Indexed pages section. Instead, those pages may be found in the Indexed pages section. Check if the pages listed in the spreadsheet are listed within the Indexed URLs section. If the page is indexed, that rules out indexing issues as a potential reason for why this page is not ranking in search results. In the example spreadsheet shown in Table 1-7, the /support/instructions page is listed within the Indexed section, even though it does not rank in search results.

Table 1-7. Example Spreadsheet to Diagnose Crawling and Indexing Issues, Step 5

Page URL	Ranking?	...	Crawl Count	Crawl Tests and Issues	Index Problems
/			76		
/products			48		
/products/blue-widget	No		5	No crawl issues detected	Duplicate without user-selected canonical
/about-us			59		
/support	No		0	Blank page and no HTML returned in the live test and headless browser—not changed with time adjustments	Discovered—currently not indexed
/support/instructions	No		20	No crawl issues detected	No issues reported—listed as indexed
/customer-login	No		25		Excluded by a "noindex" tag
/blog	No		0		Blocked by robots.txt
/blog/a-great-post/	No		0		Blocked by robots.txt

Finally, it is important to remember that Google and other search engines are not making arbitrary decisions about what pages to include in the index. Instead, these decisions are based around which pages people want to see in search results. Because of this, another means of understanding why a website's pages may not be indexed is by reviewing user engagement metrics. For example, in a web analytics tool, check the engagement or conversion rates for any page listed on the spreadsheet that is not ranking in search results. If the engagement or conversion rates are low on particular pages, that indicates people might not be interested in the page—possibly because of content-related issues or other errors present on the page. Robots are not using a website's analytics data directly to make indexing decisions but can pick up on other signals that also show people might not be interested in a page. If robots detect a lack of interest in a page, that could prevent the page from being indexed and ranked.

Step 6: Is There a Crawl, Index, or Rank Problem?

By this point of the analysis, it should be clear if a problem exists with crawling or indexing. Are robots crawling the page? Can robots crawl the page successfully? Are robots crawling successfully, but making decisions to not index the page?

In reviewing the example spreadsheet assembled over the last several steps, shown in Table 1-8, we now can understand why certain pages are not ranking.

Table 1-8. Example Spreadsheet to Diagnose Crawling and Indexing Issues, Final State

Page URL	Ranking?	Noindex	Disallow	Crawl Count	Crawl Tests and Issues	Index Problems
/				76		
/products				48		
/products/ blue-widget	No			5	No crawl issues detected	Duplicate without user-selected canonical
/about-us				59		
/support	No			0	Blank page and no HTML returned in the live test and headless browser—not changed with time adjustments	Discovered— currently not indexed
/support/ instructions	No			20	No crawl issues detected	No issues reported— listed as indexed
/customer-login	No	noindex		25		Excluded by a "noindex" tag
/blog	No		disallow	0		Blocked by robots.txt
/blog/a-great-post/	No		disallow	0		Blocked by robots.txt

To review and recap

- **/products/blue-widget**: Does not rank because of an indexing issue resulting from duplication. By resolving the duplication, this page should begin to rank.

- **/support**: Does not rank because of a crawl issue, likely due to a rendering problem. This should be reviewed more deeply, and alternative rendering methods should be tested.

- **/support/instructions**: Does not rank because of some reason other than crawling or indexing issues—such as problems with the content itself, internal links (orphaning) or backlinks. This needs to be investigated further.

- **/customer-login**: Does not rank because of the noindex directive. By removing that directive, the page should begin ranking (assuming no other issues are present).

- **/blog and /blog/a-great-post/**: Do not rank because of the robots.txt directive. Removing that directive will allow robots to crawl, and that should lead these pages to rank (assuming no other issues are present).

There is rarely a simple, sitewide answer for why pages are not ranking in search results. Understanding why pages do not rank requires investigating each page. This investigation is less about completing certain tasks—instead, it is about asking questions to understand how robots are handling each page. Answering those questions will clarify the critical issues that are holding back rankings—whether the issues are related to crawling and indexing or related to other issues present on the website.

Measuring and Monitoring Guidelines: Crawling and Indexing

- Monitor the website's log files on a regular basis (weekly or monthly for active websites, quarterly or semiannually for less active websites).

  - Track which pages are being crawled. The pages crawled will not correlate perfectly with rankings or performance, but these pages do indicate where robots are focused and what pages will likely be included in the evaluation of the website.

- Using data from the log files, determine if robots are focused on the correct pages of the website. If robots are not focused on the correct pages or are not crawling important pages, that can represent a crawling problem. See the "In Practice" section for more details on how to detect and handle these types of problems.

- Review disallow and noindex directives used on the website and evaluate if these directives are working properly and as expected. As part of this review, check if any disallow and noindex directives are conflicting with each other. As well, check if any disallow statements prevent robots from crawling JavaScript, CSS, image, or video files. Remember:

  - A disallow on the robots.txt file tells a robot not to crawl a page, file, or directory.

  - Meta robots noindex tells a robot not to index the page.

  - Meta robots and X-Robots nofollow tell a robot not to crawl links contained on that page or file.

- Check if any links contained on the website need to use link qualifiers (nofollow, sponsored, ugc) given the nature of the link. If rel qualifiers are already in use, confirm those link qualifiers are used correctly on links (e.g., qualifiers should not be included for internal links).

- Check that staging or development environments are restricted from public access, including access by search engine robots. A disallow or noindex directive is not enough to prevent robots from accessing these environments.

URL and Domain Structure

This chapter covers proper URL structures and the way URL structures can affect SEO. Search engines use a page's URL as a unique identifier for each page. That means every signal a search engine finds about a page is attached to that distinct URL. However, there are many possible URL variations, especially where parameters are concerned, and those different variations can lead search engines to identify one page under multiple URLs. This creates problems that disrupt SEO performance. A key step in improving a website's SEO performance is making sure that URLs are properly structured so that search engines can crawl, index, and rank each page correctly.

What Are the Parts of a URL?

A URL contains several parts. Using this example URL, Table 2-1 reviews each part:

```
https://go.site.com/my-page/?a=1234
```

© Matthew Edgar 2023
M. Edgar, *Tech SEO Guide*, https://doi.org/10.1007/978-1-4842-9054-5_2

Table 2-1. Parts of a URL

Item	In This Example...	Notes
Protocol	https://	For a website URL, this is either http:// (nonsecure) or https:// (secure).
Subdomain	go	The term "subdomain" can also refer to the hostname containing the subdomain (go.site.com in this example). Also, note that www is considered a subdomain, but it is so common that it is created by default unlike other subdomains.
Domain	site.com	This can also be referred to as a root domain.
Top-Level Domain (TLD)	.com	Other examples include .org, .net, or .edu. A country-specific Top-Level Domain is called a Country Code Top-Level Domain or ccTLD (like .ca for Canada or .in for India).
Hostname	go.site.com	www.site.com would be another example of a hostname. The hostname is the subdomain + domain.
Path	my-page	Another example would be /my-page/my-page-2.html. The path can also be referred to as a directory, subdirectory, folder, subfolder, slug, or file location.
Query String	a=1234	Also called a query component.
Parameter	a	The part of the query string before the equal sign. Multiple parameters are separated with an & symbol (such as a=1234&b=5678).
Parameter Value	1234	The part of the query string after the equal sign.

What Is a Canonical URL?

A canonical URL is the official, or preferred, version of a URL. A canonical URL is important to define when more than one URL can return the same content. For instance, if a page can be reached at either asite.com/page or asite.com/page.php and both URLs return exactly the same content, this creates duplication and risks confusing robot and human visitors. Defining the canonical URL can help to alleviate some of this confusion (see how to define a canonical URL in Chapter 3).

What Is a Canonical Domain?

A canonical domain is the same concept as a canonical URL but applies at the domain level—it is the official version of how the domain is presented.

There are four variants for a domain:

- `http://asite.com`
- `http://www.asite.com`
- `https://asite.com`
- `https://www.asite.com`

These variants can be seen more clearly in a grid, as seen in Table 2-2.

Table 2-2. Domain Variants

	HTTP	HTTPS
WWW	`http://www.asite.com`	`https://www.asite.com`
Non-WWW	`http://asite.com`	`https://asite.com`

A single variant should be selected as the canonical domain.

Why Select a Canonical Domain?

Each domain variant will be treated as a unique website, creating four separate websites for each domain variant. In some cases, multiple variants can be indexed in search results, resulting in lower rankings and lower performance. Selecting a single variant as the canonical domain can avoid these issues. There are two primary questions to answer when selecting a canonical domain:

- Should the canonical domain use www or non-www?
- What protocol should the canonical domain use (HTTPS or HTTP)?

HTTPS Protocol and SSL Certificates

An SSL certificate encrypts information about a visitor's session on a website, which makes that session more secure. As of 2014, using an SSL certificate is a ranking factor for Google, which means having an SSL will likely help a website rank higher in search results. Along with the SEO benefits, an SSL can also help increase trust in the website (and trust in the company managing the website), which can help increase how many human visitors engage and convert on the website. For all these reasons, it is best practice to use an SSL certificate and select a canonical version of the domain that includes the HTTPS protocol.

Should My Domain Use WWW or Non-WWW?

Whether the canonical domain includes "www" is largely a matter of choice, with no universal, measurable impact on search engine performance. Different customer segments and different companies will have different preferences for using a www in the domain. It is important to pick one version, using that version consistently and indefinitely.

Enforcing a Canonical Domain

Even if the https/www version of a domain is chosen as the canonical domain, visitors may still attempt to access other versions of the domain. All noncanonical versions of the domain need to redirect to the canonical version (see Chapter 8 for more information about redirects). For example, if https/www is chosen as the canonical domain, the http/www, http/non-www, and https/non-www variants would all need to redirect. These redirects make it clear to robots which version of the domain is the canonical version.

Along with redirecting variants to the canonical version, all links within a website (and any external links within an organization's control) should reference the canonical domain. As well, all links listed in an XML sitemap (see Chapter 5) should use the canonical domain. Consistently linking to the selected canonical domain avoids sending mixed signals about which version of the domain is preferred.

Should a Website's Pages Be Located on a Subdomain?

A subdomain has an equal chance of ranking (or not) in search results as any other part of a root domain. Remember, www and non-www are effectively subdomains too. However, search engines usually view subdomains as separate and distinct websites (this.site.com is distinct from www.site.com). As a result, any positive signals (like many links from other websites or higher-quality assessments of the content) that may help the subdomain this.site.com rank will likely have little to no benefit for www.site.com (or the reverse). Those same positive signals referencing a subfolder, though, will help all pages on that folder's subdomain. For example, many external websites linking to www.site.com/that, would typically also benefit other pages on www.site.com.

The best practice is to use subdomains only when the content on a subdomain is wholly separate from the content on the main domain and no positive signals need to be shared between content. If the content on a subdomain is not separate from the content on the main domain, then the content should typically be organized within the main domain instead. For example, a blog is often connected to the content on a main website, so a company's blog should be located in a subfolder, such as site.com/blog, instead of on a subdomain, such as blog.site.com. This way, any positive signals referencing blog posts also benefit the main domain. However, the content that is separate from the main domain, like support content provided by an independently operated subdivision or by a third-party partner, might make more sense to host on a subdomain instead of incorporated within the main domain. Subdomains also make sense for content that targets a separate audience, like a translated version of the website that targets an audience speaking a different language.

Should Individual Pages Be in the Root Directory or in Subfolders?

An example of a URL in the root directory would be site.com/some-blog-post. An example of a URL located in a subfolder would be site.com/blog/some-blog-post. Both URLs have an equal chance of ranking; however, the subfolder can sometimes help to better explain how the page fits into the rest of the site. In this example, containing the URL in a /blog/ subfolder makes it clearer this page is part of the website's blog and that may help robots (and human visitors) better understand the website's structure. Organizing pages in folders can also help with reporting. However, the benefits to where a page ranks in search results or how many clicks the page will get from a search result listing are quite minimal.

Should Individual Pages Be a File or a Subfolder?

An individual page on a website can be located at a file (site.com/page.html) or a subfolder (site.com/page). Both approaches have an equal chance of ranking and, once ranked in a search result, an equal chance of being clicked on by the people conducting the search. This is a matter of choice and usually determined by the underlying technology powering the website.

What Is Excessive Folderization?

Excessive folderization is the practice of putting pages into multiple subfolders to get more keywords into a URL, for example, site.com/blog/clothing/fashion/latest-trends/blog-post. This practice is done in the hopes the page will rank better or in the hopes people will be more likely to click that URL. However, these URLs tend to look spammy or manipulative and often do not provide much, if any, benefit. Unless technically required, it is best practice to keep folderization at a minimum.

Optimizing a Page's URL

Whether a subfolder or a file, the individual page's path should use descriptive words. For example, an about page's URL of /about-acme-corp is better than simply /about, and both are better than a randomly generated alphanumeric identifier URL, such as /3d1s7y.

However, individual URLs should not be overly long or attempt to stuff in keywords, such as the URL /about-the-best-new-york-law-firm-nyc-acme-corp-located-in-new-york-city-near-you. Long URLs run the risk of robots flagging the page as spammy or manipulative and run the risk of human visitors being put off by the URL and not clicking it.

Using a Trailing Slash

If a page's URL is located in a subfolder instead of a file, it will not have an extension (such as .html). In these cases, the next question to consider is if the page's URL should end in a trailing slash. As an example, a page's URL could be site.com/z or site.com/z/.

There is generally no difference in search performance for pages with or without the trailing slash. However, allowing both variations to return content presents a potential duplicate content issue if the slash and nonslash variants are both crawled or indexed by robots. As well, allowing both URLs means a robot can crawl two URLs instead of one, wasting crawl budget.

The best solution is to redirect to the preferred version of the page's URL. In the preceding example, site.com/z could redirect to site.com/z/ (see Chapter 8 for more about redirects). Alternatively, the problems can be resolved by defining a canonical to the preferred version of the URL (see Chapter 3).

Duplication Resulting from Query Strings

Query string usage can result in multiple URLs returning the same content. For example, site.com/my-page and site.com/my-page?p=4 are distinct URLs but could return a page containing identical content.

Tracking parameters are the most common example of query strings that do not alter a page's content. For instance, a query string could contain Facebook's tracking parameter of fbclid. Including this parameter in the URL would not (and should not) change the page's content. Instead of changing the page's content, these parameters are added to a URL so that analytics programs can monitor which visitors came to the website from a particular source. Duplication can be avoided for these types of query strings by defining canonical URLs and setting canonical tags appropriately (see Chapter 3 for additional details).

Other query strings can change the content of the page, such as a parameter that filters the page or a parameter that sorts the information presented on a page. In these cases, the URL with the query string loads distinctive content, making it a unique page on the website. Simply because it is a unique page does not mean that URLs with this query string ought to rank in search results. Instead, the question is: does the query string change the page's content in a meaningful enough way to make it worth ranking in search results? Two examples will help illustrate this point.

- **Example #1**: A "sort" parameter is added to a directory page's URL and, once added, changes the order of the items listed in the directory to either list the content alphabetically or reverse-alphabetically. The sorting functionality is likely meaningful for visitors once they arrive on the page, but the way the page is sorted would not change how the page should rank in search results. Whether the URL in search results is sorted alphabetically or some other way does not determine if people will click from the search result to this page.

  As well, the URL with the alphabetical sort, the URL with the reverse alphabetical sort, and the original URL without any sort parameter added would all return pages that contain essentially the same content—just content sorted slightly different. This means these pages duplicate each other and would compete with each other for the same search rankings.

  A canonical URL should be declared selecting one of these URLs as the preferred version to avoid the duplication and the self-competition.

- **Example #2**: A "color" parameter filters a product category page to only show products of a particular color. This type of filter would create multiple URLs— one for the original category page unfiltered plus one version of the filtered category page for each color provided. In this example, however, the color filter is meaningful to visitors on the website and to visitors arriving on the website from a search result—people looking for red products would want to see a page containing red products in an organic search result.

  As well, there would not be the same level of competition between these pages. Unlike the first example, there would likely be different search terms each of these pages could rank for, with terms for red products, blue products, green products, and all other color options.

  As a result, each page should declare itself as a unique page by declaring a self-referencing canonical—the unfiltered page should canonical to the unfiltered page. In this example, the URL with the red color filter would canonical to URL with the red color filter, and so forth.

Duplication from Multiple Query Strings

Duplication can also occur if there are multiple parameters used and the order is not fixed. For example, these two URLs set the same parameter values but in a different order. Since the values are the same, the URLs would return the same content, creating duplicate content.

```
site.com/shop?id=4&loc=567
site.com/shop?loc=567&id=4
```

The best way to address these problems with query strings is to establish clear rules for how to link to pages within the website. In the preceding example, the rule to avoid duplication resulting from multiple parameter usage may be that the "id" parameter appears first and "loc" second (or, more simply stated, parameters should be shown in alphabetical order).

Along with linking correctly internally, it is also helpful if the server can control the parameter order with redirects. For example, if a visitor requests a URL with "loc" first and "id" second, the server could redirect to the URL with "id" first and "loc" second (see Chapter 8 for more about redirects).

In Practice: Merging a Blog Subdomain

Hosting a blog located on a subdomain is common, particularly on larger or enterprise websites. As an example, the blog with all the related blog content is located at blog.site.com, while the main website is located at `www.site.com`.

This configuration is typically due to technical constraints. The blog might be managed by a different platform than the main website. If the two different platforms conflict, it can be easier to host the blog's platform in a separate environment on a subdomain instead of hosting it alongside the platform that powers the main website. With some platforms, it may not even be possible to host the blog and the main website in the same environment if there are vastly different server configuration requirements.

In other cases, there are business constraints that necessitate the subdomain. For example, the blog is set up by an external, third-party developer instead of the in-house development team because the in-house team simply does not have the resources available to develop the blog. It is easier, and more secure, to grant an external developer access to a subdomain instead of the main website.

Regardless of why the blog subdomain exists, it can present a problem for SEO because this separates the blog content from the main website's content. As discussed earlier in the chapter, a subdomain is treated as a separate website, and any positive signals associated with the subdomain will not transfer (as much, if at all) to the main website. Because of this, many companies wonder if merging the blog subdomain into the main domain by migrating the blog subdomain to a blog subfolder (from blog.site.com to `www.site.com/blog`) would result in better SEO performance.

Merging can have benefits by combining the blog content with the content of the main website, ideally boosting the performance of both. That is especially true if the current blog subdomain has decent performance, if not great performance. However, this type of move is also highly disruptive because every URL on the blog is changing. Following the merger, robots would have to find all of those new URLs, then update the index and rankings accordingly. That process of updating URLs can take weeks or months to fully complete. Rankings can drop during that time—and sometimes those rankings will not return. As a result, the disruption from the URL change can affect overall performance, resulting in larger and persistent drops in traffic.

The question, then, is should the blog subdomain be merged and, if so, how? Of course, while this example focuses on a blog subdomain, these questions still apply for other types of subdomains as well.

Questions and Considerations

Review Existing Constraints

The first question to address is if the merger is possible given technical or other constraints. To begin, the most common technical constraint is when the blog is hosted on a subdomain because the blog's hosting requirements are incompatible with the main website's hosting requirements. This means the blog and the main website cannot be hosted in the same environment. The best-case scenario is to find that the technical incompatibilities were a relic of the past and no longer exist. In that scenario, it would now be possible to host the blog in the same environment as the main website. The migration, technically at least, is more straightforward: move the system managing the blog, along with the associated files and database, over to the main website's hosting environment in a blog subdirectory.

More likely, the incompatible hosting requirements will still exist. In that situation, it may not be technically possible to host the blog in the same environment as the main website, but it may be possible to set up reverse DNS (rDNS) for the blog instead. That way, the blog can exist within a directory of the main domain (`www.site.com/blog`) even though there are separate hosting environments used for the blog and the main website.

If rDNS is not possible, another option is to configure a method of managing the blog within the main website's content management system. For example, if the blog is currently hosted on a subdomain and is powered by Squarespace, then it cannot be hosted within the same environment as the main website, and rDNS would not be an option. That would prevent the subdomain merger unless the blog is moved out of Squarespace. The question, then, is how to build the new management tool for the blog?

If the main website is managed by a custom-built content management system, then a blog management system could be built out from scratch within that custom-built system. If the main website uses WordPress, the process is much easier, and the blog simply needs to be enabled. Once configured, either by enabling a built-in feature or building a new system from scratch, the blog content would need to be moved from the old system (Squarespace in this example) to the new system. This is often a more complicated solution.

Along with technical constraints, there can also be business constraints that resulted in the blog being hosted on the subdomain. The solutions for these constraints vary depending on the nature of the business and quickly expand outside the scope of this book. However, the solution ultimately requires either making resources available internally to host and manage the blog on the main website *or* trusting a third-party developer with access to the main website.

There is often a mixture of technical and business constraints that result in the blog being hosted on a subdomain. Before moving forward with merging the blog into the main website, these constraints must be fully understood and addressed. If the constraints cannot be resolved, then none of the other questions in this section matter.

Changing Blog Platforms

Any changes made to the platform managing the blog as part of the migration can further complicate the subdomain merger. The new platform might output the blog posts using a different HTML template that changes how bots have to process the text and images. Or, the new platform might render the blog posts via JavaScript instead of in the server-side rendered code (see rendering in Chapter 3). As another example, the blog's organizational structure might change if the new platform does not support tags or categorization in the same manner as the current platform. There are plenty of other differences to consider, such as changes with how comments are managed, how videos are embedded, how images are included, or how schema is added to the page. The new blog platform used post merger might also slow load times for the blog compared to the load times for the subdomain.

As mentioned before, merging a blog subdomain to the main website is already a change because every URL changes. Any additional changes beyond the URL change increase that disruption and increase the chances of negatively affecting the performance of the blog posts following the merger. Some changes are unavoidable (like changing URLs), but if there are too many technical changes, the resulting disruption and traffic hit can make it better to forgo the merger and leave the blog hosted on a subdomain. At a minimum, all of these changes need to be carefully documented prior to the subdomain merge. As best as possible, the changes should be minimized, and avoid any changes that will make the blog worse—like a change that requires removing schema, comments, or some other important element.

Reviewing Blog Performance

There is an assumption baked into most subdomain merger projects: by merging the subdomain into the main website, there will be a corresponding boost in SEO performance for the main website and in the performance of the content currently on the subdomain. The content will be better together than apart.

However, this is only true under certain circumstances, and the performance needs to be well understood to determine if the merger makes sense. At a minimum, the following metrics should be reviewed to gauge overall performance:

- **Organic traffic levels**: How many visitors arrive on the main website and the blog subdomain from organic search results? For the sake of this assessment, traffic from other sources, like paid search, are not relevant to determine organic search performance. However, those other traffic sources could be reviewed for additional information about how the subdomain and main website perform.

- **Rankings**: Do the main website and blog subdomain rank highly in competitive search results?

- **Crawl volume**: Do robots crawl the content on the main website and blog subdomain regularly?

- **Backlinks**: Are there lots of high-quality websites linking to the main website and blog subdomain?

- **Conversions**: How many visitors engage and convert on the main website and blog subdomain content?

These metrics will help clarify if the pages on the subdomain or the main website are performing well.

If the pages on the subdomain (blog posts in this example) are not performing well across the majority of these metrics, then merging in the subdomain will not help the main website's performance—the blog does not have any performance and, therefore, cannot provide a performance boost to the main website. Similarly, if the main website is not performing well, then merging in the blog subdomain's content will not boost the blog content's performance—the main website does not have any performance and cannot provide a boost to the blog content. Merging the subdomain into the main website when overall performance is low would not necessarily hurt, but it may not be worth the effort either because traffic following the merger would remain about the same.

If both the subdomain and main website are performing at an average level, then combining could improve the performance of both; merging two mediocre sets of content can result in one good set of content that has a chance to perform better. This is a good reason to merge subdomains, and following the merger, traffic levels should increase, at least slightly.

If the main website's content is performing well while the subdomain's content is not, then merging in the subdomain could help the content on the subdomain perform better with little to no risk to the main website's content. This performance scenario supports merging the subdomain into the main website with little overall risk to the subdomain's content. Following the merger, the subdomain's content should see an increase in traffic.

Finally, if the blog subdomain is performing strongly at present, then merging it into the main website may boost the main website's performance, but only by harming the blog content's performance after the merger. This happens because rankings drop while robots are updating the index to the new URL (from blog.site.com to site.com/blog), but robots do not always reinstate the rankings to the premerger levels. For some companies, seeing a drop to the blog content's performance but a gain on the main website's performance is an acceptable outcome. This is because the main website's content leads to more conversions making that content more important for overall business objectives. For other companies, it would be unacceptable to see a drop in performance for the blog content following the subdomain merger, even if there is a corresponding gain in performance for the main website content, because there are too many conversions happening from blog content. Depending on the business's goals, this performance scenario can result in a higher-risk subdomain merger.

How Related Is the Content?

Another assumption behind a subdomain merger is that the subdomain's content fits within the main website. By merging the blog into the main website, human and robot visitors alike will have a better experience visiting the website and reading the blog posts. However, before moving forward with the merger, that assumption should be validated.

In most cases, the blog posts on a blog subdomain are often highly related to the content contained on the main website. The main website might offer ways for customers to purchase the company's products and services while the blog posts provide more details about those products and services. Merging these blog posts into the main website would make sense for visitors.

In other cases, the subdomain's content might not fit into the rest of the website. As an example, an organization's blog posts might be written primarily in the first person by the founder covering topics only occasionally related to the company's core services, while the main website is written more neutrally with content focused specifically on the company's core services. Containing both sets of content on the same website might not make sense for visitors. It might end up creating a jarring experience to move between blog posts and the main website because of the difference in writing style or the structure of the different content. In this example, then, the blog operates more like a separate website and probably should remain a subdomain.

While the content fit can be assessed qualitatively, this can also be reviewed quantitatively. One key metric is the count of internal links between the main website and the subdomain. This count can be obtained by crawling the website in a tool like Screaming Frog. To get a more accurate count of internal links, do not include any internal links contained in any of the website's navigation (header, footer, sidebar, or breadcrumb). A link to the blog subdomain will likely be found in the navigation, but to assess content fit, it is important to see if any pages on the main website link to the blog subdomain within the page's main content. If there are a lot of internal links on the main website referencing the blog and a lot of internal links on the blog linking to the main website, that indicates a deeper relationship exists between the main website and subdomain. This relationship suggests people and robots see a connection between the subdomain's content and the main website's content already.

If there are internal links, the next question to ask is if visitors use those internal links. Using a web analytics tool, determine how many visitors click on links from the blog posts to the main website and from the main website to blog posts. For example, this can be done with event tracking in Google Analytics. The more visitors who use those internal links, the stronger the relationship is between the main website and the blog subdomain.

The more internal links exist and the more those internal links are used, the greater the chances people (and robots) view the blog subdomain as part of the main website already. A merger would only help solidify that view. However, if no internal links exist, and there is no natural way to add internal links, or people are not using internal links that do exist, that suggests the subdomain should remain separate from the main website.

Completing the Merger

If the various constraints can be overcome, the technical changes can be minimized, the performance gains are worthwhile, and the subdomain's content fits appropriately on the main website, then it makes sense to move forward with merging the subdomain into the main website. This section discusses the steps needed to successfully complete the merger and minimize negative SEO impact.

Keep in mind that after considering all the various factors discussed in the previous section, it may not make sense to proceed with the subdomain merger. Inaction is an appropriate outcome and often the best decision to maintain SEO performance.

Steps 1–3: Preparation

Step 1: List All Subdomain Files

The first step in merging a subdomain is compiling a complete list of every file on the blog subdomain. This is more than simply listing each blog post but includes every other file type on the blog that contains content visitors and robots might want to see. Typical file types include

- Main blog page
- Blog posts
- Comment pages
- Category pages
- Tag pages
- Author pages
- Pagination series
- Images
- Supporting files that can be downloaded (PDF, DOC, XLS, etc.)

This list of files can usually be compiled by extracting this information from the database containing all the blog's content or by listing every image contained in an uploads or media directory. Internal resources will provide the most reliable list of pages. However, if the list of files cannot be compiled by internal resources, the pages can also be obtained by using a crawl tool like Screaming Frog, which will review which pages have received traffic, or by pulling a list of pages with external links.

Step 2: Collect All Data About Each Page

Next, the details need to be collected about each page listed in Step 1. This includes

- Current URL
- Title
- Meta description
- Meta robots
- Canonical
- The full HTML of the page's main content (including header tags, image alt text, etc.)

- Comments associated with each blog post
- Hierarchy associated with each blog post (category, tag, author)

All of these details eventually will need to be transferred over to the new blog pages on the main website, but at this step, it is about knowing what all of those details are. That way, when it comes time to move the pages over to the main website, you can confirm everything important was moved over correctly. The goal of the merger is to change as little as possible to avoid too many disruptions.

Once this data has been collected, it is typically best to freeze any changes to the subdomain's content. If any changes are made to the subdomain's content, then these details will need to be updated.

Step 3: Establish New URL Structures

The next step is determining what the URL structure should be for each file type contained within the blog following the merger. As an example, say the new blog will exist in a /blog/ directory on the main website, so the blog's home page URL will change from blog.site.com to site.com/blog/. For every other file type, what is the current URL structure on the subdomain and what should the URL structure be once the subdomain's content is merged into the main website?

A merger like this presents an opportunity to change any existing URL structures. The URL will be changing anyway as it moves from the subdomain to the main website. As an example, consider blog posts that contain date subfolders in the URL structure, such as the URL blog.site.com/2022/11/15/some-blog-post. If that blog post could still be relevant next year, having the year the blog post was published in the URL may make it look incorrectly outdated. To combat this, it could make sense to remove the date from the post URL and make the new URL structure for blog posts as site.com/blog/some-blog-post instead.

In other cases, the technical constraints from the merger necessitate URL changes. If the blog will be managed in a new system on the main website, that new system might require blog posts to use a different URL structure. As the new environment is established on the main website to manage the blog, it is important to fully understand these types of technical requirements.

Once decisions are made, the system managing the blog on the main website needs to be updated accordingly to handle URLs using the new structure. These decisions will also factor into redirecting from the subdomain to the main website—see Step 5.

Steps 4–6: Merge

The following steps need to be completed at essentially the same time to ensure that visitors and robots correctly understand the subdomain merger. The more time between these Steps, the greater the chances robots will misunderstand the merger and not crawl, index, or rank the new blog content contained on the main website correctly.

Step 4: Move Subdomain Content to New Website

To begin the merger, all the files listed in Step 1 that are on the blog subdomain need to be moved over to the blog directory on the main website. This includes the blog posts themselves, the tag and category pages creating a hierarchy on the blog, and also all the images associated with the blog posts. The blog content needs to exist at the new URL structures established in Step 3, and each file should contain all the details collected in Step 2. Typically, this can be done in bulk programmatically by copying over the database and all associated template files.

To verify the merger was successful, step back through all the details collected in Step 2 and verify everything now exists on the main website. Along with checking for blog pages, check for images and any supporting files as well.

Step 5: Add Redirects and Update Previous Redirects

Next, redirects need to be added to reroute human and robot visitors from the blog subdomain to the blog directory on the main website. The redirects need to return a 301 response status code. See Chapter 8 about redirects for more details.

These redirects can often be written to operate in bulk instead of redirecting each URL individually. Bulk redirects work by matching a pattern for the old URL and assigning a pattern for the new URL.

For example, this code from the .htaccess file would bulk redirect any URL on the blog subdomain over to the blog directory on the main website:

```
RewriteEngine On
RewriteRule ^(.*)$ https://www.site.com/blog/$1 [L,R=301]
```

To take a more complex example, say in Step 3, a decision was made to change the URL structure for blog post URLs to no longer use date subfolders. If the blog post URL on the subdomain creates URLs like /2022/11/15/blog-post-name-here/ and on the main website, after migration, the URL should be /blog/blog-post-name-here/. The following code in the .htaccess file could handle this pattern and drop the dates from all blog posts in bulk:

```
RewriteEngine On
RewriteRule ^([0-9]{4})\/([0-9]{2})\/([0-9]{2})\/([a-z\-]+)\/$ https://www.
site.com/blog/$4 [L,R=301]
```

Not all redirects can be handled in bulk. Some URL changes made in Step 3 might prevent pattern matching and negate bulk redirects. In that case, individual redirects would need to be established instead.

Remember that images need to be redirected as well. Images can account for a sizable amount of traffic for blogs, and redirecting images to the new location can help preserve traffic. Also, human visitors may have bookmarked important images, like infographics, so redirecting images can help these visitors find the new location of those images.

Along with adding in the new redirects to take robot and human visitors to the new blog URLs, existing redirects should be reviewed as well. See Chapter 8 for more details about updating redirects.

Note   The step is listed here because the redirects should be added only after the subdomain's content is moved over. However, redirects can often be written in advance of the merger during Step 3.

Step 6: Update Internal Links and XML Sitemap

Following the merger, the blog on the main website has replaced the subdomain as the canonical version of the blog. Similar to enforcing the canonical domain discussed earlier in this chapter, steps need to be taken after the merger is complete to enforce the canonical URL of the blog. This includes updating all references in internal links, the XML sitemap, and within the code to use the new blog URLs instead of using the old, subdomain URLs for the blog.

Following the merger, the entire website should be crawled to identify any internal links referencing the blog subdomain. Each of these URLs should be updated. Often, this can be updated in bulk with a find and replace command in the database. Along with looking in content for links, HTML templates and JavaScript files need to be reviewed and updated as well. Externally, search through social media or local profiles to find any references to the blog subdomain and update accordingly.

Step 7: Measure and Monitor

Once the merger is complete, the first major objective is ensuring search robots crawl the redirects and arrive at the new blog URLs. The best way to monitor this is by reviewing the website's log files to confirm Googlebot (and others) has hit the redirects and the URLs for the blog on the main website.

This can also be monitored in Google Search Console to confirm Googlebot has detected the redirects and found the new URLs.

If the redirects and new blog URLs have not been crawled within the first days following the merger, then something is likely wrong. Review the robot control settings discussed in Chapter 1 to confirm robots are able to crawl the website successfully. Also, review to make sure the redirects are properly established and that the new blog subdomain is properly enforced (see Step 6).

The next objective is ensuring the URL presented in search result rankings is updated from the subdomain URL to the URL located on the main website. Along with manually checking search results, this can also be monitored in third-party rank tracking tools or in Google Search Console. It can take several weeks to months for all URLs to fully update. However, top pages should update within a few days to a week of the merger.

The final objective is ensuring traffic returns to anticipated levels. What those levels are depends on expectations established prior to the merger. Based on the blog's performance prior to the merger, was traffic anticipated to increase, decrease, or remain about the same? It can take several weeks, or in some cases several months, for traffic levels to reset after the merger. Following the merger, continue to update blog content, update internal links, add external links, and continue to fix any technical issues that might be present. Continued optimization will help in general with SEO performance but can also help resolve any issues exposed by the merger.

Most importantly, following the merger, do not make more dramatic changes to the URL structure. A subdomain merger like this is a big change because every URL for every file on the subdomain has changed. It takes time for robots to recrawl those URLs, update the index, and update rankings. Making more URL changes will only cause further disruption to this process.

Measuring and Monitoring Guidelines: URL and Domain Structure

- Review the pages on the website. Are any pages available at multiple URLs? If so, is it clear which URL is the canonical, or preferred, URL for that page?

- Review which URLs are getting crawled by search robots on the website. Are noncanonical pages getting crawled? Are robots crawling parameter pages excessively? The more these pages are crawled, the more important it is to determine and define canonicals.

- Review subdomain usage to determine if subdomains are used appropriately and ranking as expected. If not, should the subdomain be merged into the main website? As well, review the website's content to identify any parts of the main website that act as a separate website already; should these parts of the website be separated into a subdomain?

- Review URLs to determine if any are overly long or stuffed with keywords. For any URLs that are, check how those pages are currently ranking. If the page is ranking well, no changes are likely needed, but if the URL is ranking poorly, alter the URL so that it is not overoptimized.

- Review query string usage. Do any query strings create duplication? If so, add appropriate canonical tags (see Chapter 3) and adjust parameter control to handle the duplication.

- Check if the website's canonical domain and canonical URL pattern has been established (http or https, with or without www, with or without trailing slash). Links and redirects should reference the canonical domain and URL pattern consistently.

Content Structure

This chapter discusses the technical aspects of managing and maintaining a website's content. Content must be properly structured for robots to understand the information contained within that content. This requires using proper HTML, JavaScript, and CSS code to display and render the content—including images and videos—in a way that robots can crawl it. If there are any issues with the website's code that disrupt how content, images, or videos are rendered, that can prevent pages from ranking in search results. Even simple mistakes within HTML can disrupt how robots process a page. In addition to improper code, there are other content-related issues, like duplication or thin content, that can also hold back rankings and SEO performance.

Title

The title is a tag located in the <head> of a page's HTML code. Each page should have one, and only one, title tag. Every page on the website should have a unique title tag. Example:

```
<title>Title Text</title>
```

The title tag is displayed to a visitor in the browser as the name of the document in the tab. If a visitor bookmarks the page, the title is used in the bookmark list. More importantly, the title tag is used by Google and Bing to understand the contents of the page, and the title is shown in search results.

M. Edgar, *Tech SEO Guide*, https://doi.org/10.1007/978-1-4842-9054-5_3

Meta Description

The meta description tag is in the <head> of a page's HTML code. There should be one, and only one, meta description tag for the page, and each page's meta description should be unique. Example:

```
<meta name="description" content="Text here." />
```

The meta description's content is not used by search engines to determine which search results a page should be included in or where a page ranks in search results. While the meta description's content attribute may be included in search results, search engines often show other text from the page that is relevant to the search query. Occasionally, though, search engines will use the meta description instead, and, if used, this can influence how many people click from the search result to the website. As a result, it is best practice to at least have a meta description tag on a website's key pages.

Header Tags

Headings and subheadings in a website's text can be contained in header tags, ranging from H1 to H6. There can be multiple headers within a page, but by convention, there is typically only one H1 (or top-level header) within a page, and it is most often located at the top of the page. Other headers should be used in a logical order to help convey the page's structure, and only content that presents the page's hierarchy should use header tags. It is debatable how much robots rely on header tags; however, header tags provide an indication of a page's structure and are also part of making a website accessible. Example:

```
<h1>Main Header of Page</h1>
<p>Intro paragraph</p>
<h2>Section Header</h2>
<p>Sub-section paragraph</p>
<h3>Sub-Header within Section</h3>
<p>Sub-section paragraph</p>
```

Image Alt Text

Robot visitors (and some human visitors) are unable to see images. This is acceptable if the image is decorative and does not add to the overall meaning of the page. However, if an image is part of the website's main content, not seeing the image will limit the robot (or human) visitor's ability to understand the page. This is especially true for links that use images as the link's anchor text—if a robot or human visitor cannot see the image within, they will not see the anchor text for the link, and they will struggle to understand what that link will do if clicked. This can cause search engines to not rank pages correctly in search results and cause confusion for human visitors who are unable to see those images.

Any image that is part of a website's main content or that acts as a link should include an "alt" attribute that describes the image. Alt text should be short and descriptive. Alt text should avoid using phrases such as "image of" or "photo of" since those are rarely helpful descriptions. As well, stuffing keywords in alt text will be ignored by search engines and not be overly helpful to humans either. Example:

```
<img src="tech-seo-guide-cover.jpg" alt="Tech SEO Guide" />
```

Decorative images should have an empty "alt" attribute:

```
<img src="decorative-photo.jpg" alt="" />
```

SVG Alt Text

For inline SVG images, alt text should be provided in a <title> tag. The <title> tag should be the first element within the <svg> tag. To ensure full browser and robot support, it is best practice to add an "id" attribute to the title tag and reference that "id" attribute's value in an "aria-labelledby" added to the <svg> tag. Example:

```
<svg    xmlns="https://www.w3.org/2000/svg"    role="img"    aria-labelledby=
"svg_1_title">
   <title id="svg_1_title">Brief Description of the Image</title>
   ...
</svg>
```

Rendering

The process of loading content into the browser is known as rendering, and rendering can happen on the server or in the browser (the client):

- **Server-side rendering**: The server pulls together all the HTML code and sends the code to the browser. The browser receives that HTML code as delivered, and it displays it as a web page, with no further work required by the browser.

- **Client-side rendering**: The browser receives JavaScript code along with HTML code from the server. The browser displays the HTML code but also executes the JavaScript code. Once executed, the JavaScript code can manipulate the page's content and design.

Many websites today use a combination of server-side and client-side rendering. Both are supported by search engine robots with varying degrees of success.

Client-Side Rendering and Robots

When crawling a website, robots can execute most JavaScript code but are not able to render content in certain ways. For example, Googlebot cannot trigger onclick events. Because of this, Googlebot would not be able to retrieve any content that only renders once a button is clicked. The same would be true of content rendered in response to onscroll events, like content that appears as part of infinite scrolling or images loaded via lazy loading.

The only way to ensure robots can retrieve all content on a web page is to load that content via server-side rendering. Server-side rendering is also simpler for robots to load, requiring less processing and ensuring better results. If client-side rendering is used, the important content that robots must retrieve in order to understand the page should render with the onload event.

What Is Dynamic Rendering?

Google supports dynamic rendering of content, though does not recommend this as a long-term solution. With dynamic rendering, JavaScript code is executed, and the resulting HTML changes are rendered in advance by using a rendering tool. This is called prerendering. Prerendering avoids the need for robots to do any of the rendering work on their end. Essentially, this process converts the client-side rendered code into server-side rendered code. This provides more control over how everything is rendered and over what robots are able to retrieve within the page, helping robots to not miss the website's important content.

What Is a Noscript Tag?

A <noscript> tag contains content visitors or robots should see if they cannot execute the JavaScript code. If a page's main content is only loaded via JavaScript, containing the main content in a <noscript> tag can help make the page more usable and accessible.

One common instance of this is with lazy-loaded images. With lazy loading, images are not loaded into the website on page load. Instead, images are only loaded when they appear on the screen after they scroll into view. This means lazy loading requires JavaScript code to detect scrolling behavior and determine when an image should be loaded into the browser based on the scrolling behavior. Because robots do not scroll, they will not trigger an onscroll event and will not be able to detect these images. Instead, the best practice is to include the image in a <noscript> tag.

In this example of a lazy-loaded image, the "src" attribute is blank until the image is scrolled into view. Because robots will not scroll and would not execute the JavaScript to trigger the lazy load behavior, the noscript tag provides the image URL, allowing robots to crawl the image.

```
<figure>
      <img src="" alt="description of image">
      <noscript><img src="path/to/my-image.png"></noscript>
</figure>
```

Pagination

Each page in a paginated series should be treated as a distinct and unique page on the website. This means, each page in the paginated series should canonical to itself, contain a unique title, and have a unique meta description. As well, each page's content should differentiate it from the other pages in the series. The URL for the pages in the paginated series should put the page number in the URL slug (such as /blog/2) or in the query string (/blog?page=2), instead of using a fragment (/blog#2).

Infinite Scrolling and Load More

Robots will not scroll down a page to trigger an infinite scroll. Similarly, robots will not click a "load more" button to trigger the load of additional content. As a result, robots will not be able to load content reached through an infinite scroll or a "load more" button. If this content is important to understand and rank a website's pages, this presents a critical problem for the website's organic search performance. The best solution, and the one recommended by Google, is to make a traditional paginated series of pages available as an alternative for robots (and for visitors who cannot use JavaScript).

What Is Duplicate Content?

Duplicate content occurs when an entire page of content or a substantial amount of a page's content is repeated on multiple URLs. While typically thought of as exact duplication, duplicate content also includes near duplication, which is when two pages address the same topic in very similar ways, though not in exactly the same way.

Duplicate content typically is contained within one domain where multiple pages on the same website present the same content. However, duplicate content can also occur across multiple domains where pages on different domains (or subdomains) contain the same content.

Why Does Duplicate Content Present a Problem?

Duplicate content (exact or near duplication) often confuses search robots as they decide which pages to rank (or not) in search results. Worse, duplicate pages on the same website might compete for the same rankings creating self-competition.

Google (and other search engines) may make a decision to rank one copy of the page instead of another, but the page chosen to rank may not be the same page the company managing the website would prefer to rank. Alternatively, Google may choose to ignore and not rank any versions of the duplicate page.

In some cases, search engines may penalize a website due to excessive amounts of duplicate content and, as a result, not rank that website in search results. This is especially true if the duplicate pages seem designed to manipulate robots.

Finally, duplicate pages can worsen the experience humans have when visiting a website. As people navigate through a website, if the same content appears on multiple pages this can cause frustration or confusion, resulting in reduced engagement and lower conversion rates.

Examples of Duplicate Content

Category pages on a blog are a common example of duplicate content. If all (or most) blog posts are added to Category A, then the page for Category A will contain the same content as the main blog page, just in a slightly different order. Fixing this type of duplication requires reworking the category to ensure it has different content than the main blog page. Typically, fixing this also requires rethinking how category pages fit within the website's hierarchy and if those category pages are necessary for visitors.

Duplicate content can also happen accidentally. This is especially true for websites managed by larger companies. When multiple people or departments are responsible for writing content, it is easy for multiple pages to be created that speak to the same topics. The best way to fix this accidental duplication is through developing editorial policies that will prevent exact or near duplicate pages from being created and published on the website.

There are also technical reasons for duplicate content. For example, a content management system (CMS) might allow a page to be returned at two different

URLs. For example, a blog post might be available at site.com/post.php?id=1 and at site.com/blog/blog-title, creating duplicate content. This type of duplication is often best fixed by adjusting the CMS to only return the page at a single URL. This type of duplication can often also be resolved with canonical URLs.

As another example, a retail website may allow for a product listing page to be sorted in various ways. When the sort is set to price, the URL may change to include price in a query string (?sort=price), and when the sort is set to popularity, the URL may change again (?sort=popular). Both URLs would return essentially the same list of products, meaning these URLs duplicate each other. Canonical URLs are also a solution for this type of duplication.

What Is a Canonical URL?

One way of resolving certain types of duplicate content is to define a canonical URL. A canonical URL is the preferred version of the URL. As an example, say the URLs site.com/A, site.com/B, and site.com/C all return the exact same content. In this example, site.com/B is defined as the canonical version. This would tell search engine robots they should ignore the duplication found on site.com/A and site.com/C, ranking site.com/B instead.

Communicating the canonical URL to search engines requires adding a canonical link element to the duplicate pages. The canonical link element is in the <head> of a page's HTML code.

If site.com/A, site.com/B, and site.com/C contain duplicate content (exact or near) and site.com/B is chosen as the canonical URL for this content, the following canonical link element would be placed in the <head> of site.com/A, site.com/B, and site.com/C:

```
<link rel="canonical" href="https://site.com/B" />
```

Canonical link elements can also be defined across domains. For example, if siteone.com/A contains the same content as sitetwo.com/A, then a canonical link element could be used to tell search engine robots which page on which domain should be considered the canonical, or official, version of the content that should be indexed.

Defining a canonical URL does nothing to resolve the problems duplicate content creates for human visitors. Defining canonicals is best for areas where the duplication does not present a problem for humans but could present a problem for robots, such as duplication created by sorting items listed on a page.

On a related note, there is also the term self-referencing canonical. As the name suggests, this is a canonical defined to reference itself. For example, https://site.com/D contains a self-referencing canonical to https://site.com/D.

Other Ways to Resolve Duplicate Content...

Redirects provide another way to resolve duplicate content. For example, if site.com/X is a duplicate of site.com/Y, then site.com/Y could be redirected to site.com/X so that only one version of the content exists on the domain. (See Chapter 8 for more about redirects.)

As well, to help human and robot visitors, links within a website should reference the canonical version of a page. If site.com/H and site.com/J contain the same content, but site.com/H is the canonical version, then all links should reference site.com/H to prevent visitors from ever reaching the duplicate version at site.com/J. Having internal links reference the canonical URL reinforces the canonical URL selection and prevents robots (and humans) from finding the duplicate versions of the page.

What Are Thin Content Pages/ Doorway Pages/Cookie-Cutter Pages/Low-Quality Content?

Thin content, autogenerated pages, cookie-cutter pages, doorway pages, and more are all forms of low-quality content. Each has a specific definition, but simply put these are any type of page on a website that does not add much, if any, value to a human visitor. As with duplicate pages, low-quality pages may be intentionally created to manipulate search results or may be automatically created by the website's code. As much as possible, this content should be deleted or a noindex directive should be used. (See Chapter 1 for more about noindex directives.)

The length of the page has nothing to do with the quality of the page. Long pages with lots of words can be considered low-quality pages if all those words do not add value (like an article "written" by a spambot that contains 1000 mostly nonsensical words). Equally, a short page can be quite valuable (like an entry in a dictionary defining a word).

Low-Quality Content Problems

Search robots generally do not rank lower-quality pages. After all, search engines would rather help human visitors find high-quality pages that are relevant to the search query. If excessive amounts of low-quality pages are found on a website, the website may also face a penalty and be removed from search results. Beyond this, low-quality pages create more pages for a robot to crawl through on a website, which means robots may spend too much time crawling low-quality pages, missing higher-quality pages.

In addition to these problems for robots, low-quality content is bad for human visitors. As people browse through a website, they want to find high-quality, relevant content. If too many pages are irrelevant or present low-quality content, human visitors may leave, reducing engagements and conversions.

What Are Orphaned Pages?

An orphaned page on a website is a page that has no internal links pointing to it. There is also the concept of a near-orphaned page, which is a page that has very few internal links pointing to it. Because there are no (or few) internal links referencing this page, the orphaned page is effectively cut off from the rest of the website, which can harm that page's ability to perform. Internal links guide search engines (and visitors) through a website, so without internal links, robots may be unable to find the page; if search engine robots cannot find the page, they will not rank it in search results. Even if found, the lack of internal links suggests the page is not very important, decreasing the page's chances of ranking.

A common mistake to fix orphaned pages is to add irrelevant internal links. Doing so can create other problems and reduce the website's overall quality. Instead, to fix orphaned pages, review those pages and determine what other pages on a website could link to the orphaned page in a relevant and high-quality manner. If there is nowhere to place a relevant link to the orphaned page, then the orphaned page might need to be removed from the website.

Multilingual Websites and Pages

A website may serve the same content in different languages. For example, a website might provide blog posts in English, French, and Spanish. Each blog post would convey the same information but be translated into different languages, and there would be a unique URL for each language's version of the post.

A link element can be used to communicate how these pages are connected. This link element should include a rel attribute of alternate and hreflang attribute specifying the language. The hreflang specifies the language in ISO 639-1 format.

In the blog post example, with posts available in French, English, and Spanish, the following link elements with rel alternate attributes would be included in the <head> of each post:

```
<link rel="alternate" hreflang="en" href="https://www.site.com/Post/" />
<link rel="alternate" hreflang="es" href="https://www.site.es/Post/" />
<link rel="alternate" hreflang="fr" href="https://www.site.fr/Post/" />
```

International Page Variations

An hreflang attribute can also specify a country. While a language can be specified without a country, a country cannot be specified within a language. A country (or regions) needs to be specified in ISO 3166-1 alpha-2 format. For example, if there is a blog post available for Spanish speakers in the United States and a different version available for Spanish speakers in Spain, the following hreflang tags could be used:

```
<link rel="alternate" href="https://www.site.es/Post/" hreflang="es-es" />
<link rel="alternate" href="https://es.site.com/Post/" hreflang="es-us" />
```

Language Defaults

The hreflang attribute can also define a default version of the page using "x-default" as the hreflang attribute's value. This is helpful to specify which variation of the page should be used if a visitor's language or region does not match a language version specified.

In the blog post example, if a visitor attempts to access the post with a browser language set to German, no post would be available. An example where the English page is set as the default:

```
<link rel="alternate" hreflang="x-default" href="https://www.site.com/Post/" />
```

In Practice: Content Assessment

Finding content-related issues—like duplication, low-quality content, or orphaned pages—is an important part of maintaining and improving a website's SEO performance. The best way to find these types of issues is by conducting a full assessment of the website's content. This section discusses what to review within that assessment to identify any content-related issues present on the website.

Step 1: Obtain a Full List of Website Page URLs

The first step is finding all of the content contained on the website. For a content assessment, that includes all text-based files, like HTML pages, PDFs, or DOCs. For simplicity, all of these text-based content files will be referred to as website pages in this section.

Typically, images or videos should not be included in this type of content assessment. This is because search engines typically evaluate images and videos as part of a page, instead of evaluating each image or video as its own distinct page. A similar content assessment could be conducted for images and videos as many of the metrics discussed in this section are applicable.

Obtaining a list of all of a website's pages requires using multiple sources. Relying on a single source provides an incomplete look and risks missing at least some of the pages. There are five recommended sources to use to obtain a full list of pages and the corresponding URLs for those pages:

1. **Export pages from internal system**: If technically possible, export a complete list of all of the pages from the content management system or database powering the website. For example, on websites using WordPress, this can be done with a plugin called "Export All URLs," though other systems have similar plugins or scripts that can be run to obtain a full list.

   This export tends to capture more pages than search engines will ever find on a website. As an example, on an ecommerce website, this export would include cart and checkout pages, which search engines should not crawl, index, or rank. However, this also means this export of URLs will provide a more complete view of the website than can be provided by external sources.

   The export will typically not include pages that are not stored in the database, such as PDF, DOC, XLS, or TXT files. If there are many of these file types on a website, then they will need to be found with other sources.

2. **XML sitemap**: The website's XML sitemap can also be a useful source for the list of page's URLs. This should mirror the pages exported from an internal system. However, the XML sitemap and an export from the internal system can sometimes contain different sets of pages, especially if the XML sitemap is generated by a different system or if the XML sitemap is manually constructed.

   As discussed in Chapter 5, the XML sitemap will not

always contain an accurate list of pages. If poorly maintained, XML sitemaps can contain undesirable pages, like pages that redirect or return an error. Redirected or error pages would be treated separately from the nonredirected and nonerror website pages.

Despite potential problems, the XML sitemap does provide a list of the pages that search robots are currently finding and crawling. That means that all of the pages listed on the XML sitemap need to be reviewed within the content assessment for a full understanding of the website's content quality.

3. **Backlink target pages**: External backlinks are one of the most common sources leading search engine robots to a website's pages. The pages linked to by an external website are referred to as backlink target pages.

A list of backlink target pages will provide a different look at the website pages than an export from an internal system or an XML sitemap. This is because backlinks might reference older pages on the website. Backlink targets might also be pages that are not part of the content management system—like stand-alone HTML pages or PDF files.

There are various tools that can find a website's backlink target pages, including Ahrefs, Semrush, Moz, Majestic, SE Ranking, and plenty more. Google Search Console and Bing Webmaster Tools also provide a list of backlink target pages. Each of these tools finds slightly different backlink targets, and each tool also reports on the target pages in different ways. The best approach is to export backlink targets from multiple sources—ideally, at least three sources.

4. **Pages with traffic**: So far, the full list of a website's pages has been sourced from internal and external sources, but what pages are people visiting on the website? The best way of obtaining that list is via a web analytics tool. Even though this assessment is for SEO, the list of visited pages should be across all traffic sources, not just organic search. Every page on the website is a potential candidate for search engines to crawl, index, or rank; therefore, every page should be reviewed.

This list of pages showing traffic will typically overlap with the list of pages obtained from internal and external sources, but there can be some differences, such as pages that use parameters. For example, there might be multiple tabs presented on a web page, and clicking a different tab surfaces different content on that page. The different content on that page is indicated by the inclusion of a parameter added to the URL (site.com/some-page?tab=2). These parameters typically would not be included by an export from a content management system, would not typically be listed in an XML sitemap, and may not have backlinks. However, the content shown on the page when the parameter is included in the URL might be important content for SEO.

For best results, traffic should be viewed over a wider time range like a month or a quarter. For deeper content assessments, like those following a traffic drop, a wider six-month time range might provide a more appropriate view.

5. **Website crawl**: Finally, a full crawl of the website should be completed so that all URLs for all pages found in the crawl can be included in the full list of website pages. A crawl will overlap with the other methods already mentioned but can still highlight important pages.

   As one example, consider pagination within a category. A single category might contain dozens of pages due to pagination (site.com/some-category, site.com/some-category/2, site.com/some-category/3, etc.). All of these would be important to review within the content assessment to see if the pages are helping or hurting the website's performance. However, these pages might not surface through the other methods discussed. Pagination might receive only a few visits, so not all of these pages would surface in an analytics tool. It is very rare for backlinks to link to paginated pages, so these pages would not surface within that. Often, pagination is removed from XML sitemaps, and because pagination is autogenerated, there may not be a database record for each paginated page. However, a crawl tool would find all the links to the paginated pages.

After exporting the lists from each of these five sources, there will be at least five different lists of URLs representing all of the website pages (this could be closer to seven or eight different lists, depending on how many backlink target lists were obtained). The final step is to combine and deduplicate each of these lists. The end result should be a complete list of every URL for every unique website page.

One problem with the consolidation of these lists is that each method discussed earlier will present the page's URLs in a different format. Some will include the protocol and domain, others will only include the domain or hostname, while others will only include the path. Creating further confusion, some of the backlink targets might use a nonsecure protocol (http://) even though the website uses a secure protocol (https://). Similarly, backlinks might use www even though the website does not (or vice versa).

To get around these URL format inconsistencies, every URL identified from every source listed earlier should be formatted the same, prior to combining the lists. It is best to simplify the URL, so that it only includes the path for each page's URL—also called a relative URL.

Once every URL has been simplified to a relative format, combine all the lists and deduplicate. Then, add the relative URL for every unique page on the website to the first column of a new spreadsheet. If subdomains are part of the assessment, include a column for the hostname as well.

By only listing relative URLs, it makes opening those URLs in the browser a somewhat slower and more tedious process, especially when evaluating hundreds or thousands of pages in the assessment. To combat the tedium, it is helpful to list the full URL that includes the protocol in another column—this is called the absolute URL. In another column, create a formula to establish the absolute version of each URL. As an example, if Column A contains the path to the website and Column B contains the hostname, then Column C might contain this formula:

```
=CONCAT(B2,A2)
```

The resulting spreadsheet would look something like Table 3-1.

Table 3-1. Content Assessment—Step 1

Relative URL	Hostname	Absolute URL
/some-page.html	www.site.com	https://www.site.com/some-page.html
/another-page	sub.site.com	https://sub.site.com/another-page
/my-category/2	www.site.com	https://www.site.com/my-category/2
/	www.site.com	https://www.site.com/
/another-page?tab=2	sub.site.com	https://sub.site.com/another-page?tab=2

Step 2: Page Information

The next step is to begin gathering information about each of the website pages. This information can be useful for grouping pages together during the assessment and automating certain aspects of the assessment. More importantly, this information will begin to expose content-related issues to address on the website.

One of the easiest ways of obtaining the information listed as follows is to crawl a full list of all the page URLs in a crawling tool. Many crawl tools allow this functionality. For example, in Screaming Frog, change Mode to List and then upload a list of all the page URLs. This list should contain all the URLs in the absolute URL column of the spreadsheet created in Step 1. After the crawl is complete, the various information noted as follows will be returned for each page:

- **HTTP status code**: The first piece of information to include is the response status code returned for each page's URL (see Appendix A). Because the page's URLs were gathered from a multitude of sources, it is possible that some of the pages are errors (status code in the 4xx class or 5xx class) or redirects (status code in the 3xx class). The content assessment should only evaluate pages with a status code of 200, which indicates a normal and functioning page on the website.

  After adding a column containing each page's status code, the list of pages to assess can be filtered to only include pages with URLs that return a status code of 200. Any URLs that return an error should be flagged to address separately (see Chapter 7). Any URLs that redirect should be reviewed as well to see if there are any issues with the redirects (see Chapter 8).

- **Title**: The page's title tag text should be included in another column. For non-HTML pages that do not have a title, this can be omitted. However, if any HTML pages are missing a title tag, that presents an opportunity to improve the content and better optimize that page by adding a title tag. This review might also surface title tags that are not well optimized, such as title tags that only contain one or two words.

  Including a column for the title can also help with basic duplicate content detection. If two or more pages contain the same title tag, or a very similar title tag, that is an indication those pages might contain duplicate content. Note that Screaming Frog also provides a hash value about each page. If two or more pages have a matching hash value, that means those pages contain exactly the same content and would be considered duplicates.

- **H1**: Similar to the title tag text, a column containing the text of the H1 tag should be included for each page (non-HTML pages, like PDFs, will not have this information). Ideally, each page will only have a single H1 tag. If multiple H1 tags are found or no H1 tags are found, there is an opportunity to improve the page text and structure by either reducing to a single H1 tag or adding an H1 tag to the page.

- **Meta description**: While meta descriptions are only occasionally used in search result listings and typically do not have any influence over ranking position, it is still helpful to make sure that each page has a unique meta description where it is used in search results. The content assessment can surface any pages missing a meta description or any pages with a duplicate meta description—both of which present a low-priority opportunity to improve page content.

- **Meta robots or X-Robots noindex status**: The noindex status from the meta robots or the X-Robots tag (see Chapter 1) helps determine the priority for any issues found in the content assessment. Because pages that contain a noindex directive will not rank in search results, any content-related issues on a page with a noindex can be considered lower priority—at least for SEO. It might make sense to address some of the content-related issues on a noindex page because those content-related issues might be negatively impacting user experience or conversions.

- **Canonical**: Next, include a column for the canonical URL that is stated on the page (if any is stated). The canonical value stated on the page should be compared against the absolute URL added to the spreadsheet in Step 1. For most pages, the absolute URL should match the canonical URL—that is, each page should typically contain a self-referencing canonical.

The key question to answer is if the canonical URL does not match the absolute URL, is there a good reason for that mismatch? It could be that the URLs do not match because of a known issue that the canonical is attempting to address. In that case, those URLs should be further evaluated to see if the canonical is addressing the problem appropriately. For example, is the canonical tag helping prevent duplicate pages from ranking in search results? Alternatively, a mismatch between the canonical and absolute URLs could indicate a mistaken canonical value that needs to be corrected.

- **Language and country**: For multilingual websites or websites targeting multiple countries, the hreflang value should also be included in the spreadsheet. Typically, it is helpful to evaluate the pages within each language and country designation separately.

  Along with helping filter the pages in the content assessment, this information should be reviewed to verify it has been stated correctly. Some pages might have ended up assigned to the wrong country or are presented in a different language than indicated by the hreflang.

- **Primary targeted keyword**: Finally, it can be helpful to include a column containing the primary keyword being targeted by each website page if this information is available. Not every SEO team selects a primary keyword for each page of the website, while other SEO teams have two or three targeted keywords, so all of those might make sense to include. The targeted keyword or keywords cannot be obtained by a crawl of the pages, but will have to be brought in from another source, such as an internal database that keeps track of targeted keywords.

  Having the targeted keyword can help highlight duplication if two or more pages are targeting the same keyword or very similar keywords. As well, in the following step, the ranking position can be changed from an overall page ranking to a keyword-specific ranking, which might be a more useful metric to judge the page's overall value.

Following this step, the spreadsheet will now have expanded to several columns with information about each page. It should look something like Table 3-2.

Table 3-2. Content Assessment—Step 2

Relative URL	Hostname	Absolute URL	HTTP Status Code	Title	H1	Meta Description	Meta Robots/X-Robots	Canonical	Language and Country	Targeted Keyword

Step 3: Page Metrics

The final step is gathering data about how each page performs. Along with exposing issues with the website's content and structure, these metrics can also establish priority for any problems uncovered during the assessment:

- **Average ranking position**: The average ranking position provides a high-level summary of the page's importance. If a page typically ranks in the first few positions within search results, then the page likely contributes more to the website's overall performance. Issues surfaced in the content assessment that affect pages with a higher average ranking position are typically more critical and will have a greater impact on SEO performance.

  The average ranking position can either be for the page as a whole or it can be narrowed to only show the average ranking position for the targeted keywords (if any targeted keywords were included in Step 2). The average ranking position can be obtained from rank tracking tools—like Semrush, Ahrefs, or Moz—or it can be obtained from Google Search Console (or Bing Webmaster Tools for Bing's organic search rankings). As all of these tools evaluate ranking position in somewhat different ways, it can be helpful to bring in the average ranking position from multiple tools to get the most accurate understanding of the page's priority.

- **Impressions**: The average ranking position might say that a particular page ranks near the top of search results, but not all top-ranking positions perform equally. To better understand priority, the total impressions should be added to the spreadsheet as well. This metric can be obtained from Google Search Console or Bing Webmaster Tools and shows how many searches a particular page appeared in for a given phrase. Impressions should be evaluated alongside ranking position to establish the importance of any issues on a given page; pages with more impressions and a higher average ranking position are more important, and any issues on these pages should be fixed quickly.

Along with helping establish priority, impressions can also highlight content issues. If there are certain pages on the website without any impressions or with very few impressions, that means those pages are not showing in search results. This could be due to issues like duplication or low-quality content. Or robots might be unable to render the page and see the page's content. Pages without any impressions or with very few impressions need to be further investigated to see what issues exist and how those issues should be fixed.

- **Organic engagement rate and conversion rates**: The next question to answer is how many visitors who arrived on any given page of the website from an organic search result engaged (or interacted) with the page or converted after arriving on that page. Engagement will be defined differently on each website and can include a wide variety of interaction types—time spent on the site, usage of different features, video plays, scrolling activity, and more. Conversions will also be defined differently depending on the nature of the website—for the sake of this assessment, it does not matter if a conversion represents completions of a lead form, clicks on affiliate links, or purchases on an ecommerce site.

  Add a column containing the organic's visitor's engagement rate and a column for organic visitor's conversion rate to the spreadsheet for each page. Along with looking at the rates for organic users specifically, these rates could also be measured for all traffic sources to provide a more holistic look at how people visiting the website engage with each page and convert.

  The engagement rate provides a good indication of the quality of the page's content. If visitors think the page is lower quality or are confused by duplication, they will not convert or engage with the page.

  As a result, any pages with lower engagement or conversion rates should be carefully reviewed to see if content issues are present. That is especially true if the page with lower engagement rate happens to also have a lower volume of impressions or lower average ranking position.

- **Internal link count**: This content assessment should also identify any pages that are orphaned—meaning there are no internal links to that page. To identify orphaned pages, the total count of internal links should be included in the spreadsheet for each page. This number can be obtained from a crawl of the website. In Screaming Frog, for example, this is reported as Inlinks.

  Any pages with zero internal links are the true orphaned pages, but there are also near-orphan pages with only a few internal links. How many internal links is a "few" depends on the website and the overall average internal link count. If the pages have an average of 50 internal links, then pages with 10 or fewer internal links would be considerably below average and would likely be considered near-orphans for that website. But if the website pages only have an average of 10 internal links, then pages with only 3 or fewer internal links might be considered near-orphan pages.

  Before adding internal links to orphan or near-orphan pages, those pages need to be reviewed further. First, ask if these pages should be kept on the website. Maybe there are so few internal links because it simply is not a good page worth keeping, and it should be removed instead. Second, check if any of these pages contain a noindex tag. If so, then adding internal links to the page would not help SEO performance. Another thing to review is the canonical URL on these pages and if the canonical tag references a different URL. If so, then this page likely is not important for SEO performance, and it would not help to add internal links to this page.

This step will add seven new columns to the spreadsheet. By the end of this step, the spreadsheet should now look like Table 3-3.

Table 3-3. Content Assessment—Step 3

Relative URL	Hostname	Absolute URL	...	Avg. Ranking Position	Impressions	Engagement Rate	Conversion Rate	Internal Link Count

Content Assessment Schedule

The content assessment includes gathering a lot of information and metrics about each page of the website. However, the most important part of the content assessment is not gathering the information or metrics—the questions raised by the information and metrics are far more important. Asking these questions regularly will help uncover any issues that exist with the website's content.

This content assessment should be part of an SEO team's regular website maintenance. The larger and more active a website, the more frequently this content assessment should be conducted to resolve any issues quickly, such as every other week or monthly. For less active websites, a full assessment of the website content should be completed quarterly or annually.

Along with reviewing and assessing content as part of regular maintenance, a full content assessment should be completed before any major changes are made to the website, including redesign or redevelopment projects. As well, if the website has recently faced a major drop in organic search traffic following an algorithm update, a content assessment can help identify potential causes and identify opportunities to correct the problem to regain traffic.

Measuring and Monitoring Guidelines: Content Structure

- Crawl the website to find all titles, meta descriptions, canonical tags, header tags, and alt text on a regular basis, such as monthly or quarterly. This should also be repeated following any major website updates to detect any new issues.

    - Confirm titles and meta descriptions are unique for each page on the website. This includes paginated pages (e.g., the title of the second page in a series might be "An Interesting Category, Page 2 or 12").

    - Confirm canonical tags state the correct URL. Confirm that any duplicate pages contain canonical tags that will sufficiently resolve the duplication. As well, confirm that canonical tags point to valid, high-quality pages on the website.

- Confirm that header tags are used in proper sequence (H1 through H6) and correctly communicate the hierarchy of each page.

- Confirm alt text within images is descriptive and used appropriately for all images that are part of the website's main content. As well, confirm alt text is not included for decorative images.

- Check if robots can load all the content contained on the website, especially the main content of each page. Along with testing the website in Google's various tools, a headless browser can also be used to test what content can be loaded via JavaScript.

- If the website is available in multiple languages or serves multiple countries, check that hreflang tags are correctly communicating regional/country and language differences. As well, confirm the foreign language pages rank correctly in international or foreign-language search results.

Schema and Structured Data Markup

This chapter covers schema markup, which is optional code that can be added to a page to better explain the purpose of the content on that page to search engine robots. As well, schema markup can help enhance rankings, leading to higher click-through rates from search results. However, schema markup needs to be added correctly and accurately reflect the content presented on the page to help—if schema markup is used incorrectly, it can be ignored or, worse, can lead to errors or manual actions.

What Is Schema Markup?

Schema markup provides a way to structure information contained on websites. Most information provided on a website is in an unstructured format—there is no way for a robot to easily know what the text on a page is. A sequence of numbers might be a price, a date, or a phone number. Adding schema markup changes the website's content from unstructured to structured. By doing so, schema markup provides more details to search

© Matthew Edgar 2023
M. Edgar, *Tech SEO Guide*, https://doi.org/10.1007/978-1-4842-9054-5_4

robots about what that content contains, allowing robots to use that information to enhance search results.

What Are Properties and Types?

Schema markup is organized around different types of items. Each item type has different properties that describe that item. For example, Person is an item type within schema, and the Person item type has a number of properties, including the person's given name (in the givenName property), family name (familyName), and email address (email).

Each item type is organized as a hierarchy moving from broader item types to more specific item types. Specific item types inherit properties from broader item types. As an example, the Article item type inherits properties from the item type of CreativeWork, and CreativeWork inherits properties from the item type of Thing. As a result, when adding Article schema markup to an article, the properties from the Article item type of articleBody or articleSection can be included as can the properties from CreativeWork, such as author or dateCreated, and the properties from Thing, including the url property.

How Does Schema Improve SEO Performance?

While schema markup is not required, it can help a website's SEO performance. The biggest advantage is that schema markup can enhance a search result listing by adding additional pieces of information, such as star ratings in the case of a review schema or FAQs for pages containing questions and answers. The more interesting a search result looks, the better the chances people will click on that search result, resulting in more traffic to the website.

More generally, robots can use schema to better understand the page's content and better understand the pieces of information contained in the page's content. By making a page easier to understand, schema markup can increase the chances that the page will appear in the proper types of search results.

Does Schema Markup Help a Page Rank?

No. Schema markup does not influence where or if a page ranks in search results.

Do Googlebot and Bingbot Have to Use Schema Markup?

Schema operates as a suggestion to Google and Bing. Search engines will not necessarily use schema to enhance a search result listing, even if the schema code is implemented properly.

Can Schema Markup Cause Manual Actions?

Improper usage of schema markup can lead to manual actions if the information represented in the schema markup differs from the information represented on the page. For example, Google may issue a manual action if the schema markup shows that a product has a four-star rating, but that same rating is not shown anywhere on the page itself. Only showing information in schema can seem like a website is trying to manipulate search robots into seeing something that normal visitors would not see.

What Schema Code to Use?

All available schema markup types can be found on schema.org. It is important to note that not all schema types listed on schema.org are currently supported by search engine robots, and not all schema types will enhance search results. Google and Bing provide details about which schema types they currently support in their documentation, including required properties for each schema type.

JSON-LD, Microdata, and RDFa

Schema markup can be written in several different formats. The most common format is JavaScript Object Notation for Linked Data, or JSON-LD. JSON-LD is separate from the rest of the content contained in the HTML code. That means the information contained in the schema markup is stated twice on the website: once in the content itself and a second time in the JSON-LD markup. Placing data in two places can sometimes lead to problems where the information provided on the website does not match what is stated in the JSON-LD schema markup.

While JSON-LD is Google's recommended format, Google and Bing do support markup in Microdata and RDFa (Resource Description Framework in Attributes) formats. Microdata and RDFa add additional attributes to the HTML tags. This means information only needs to be stated once on the website instead of twice because the HTML containing the information can be adjusted to include the necessary schema attributes. Microdata and RDFa are not as robust or expressive as JSON-LD, limiting opportunities when using those formats.

In Practice: Writing Schema

There are tools that will dynamically generate schema and ways to automate schema generation within content management systems. However, the best way to understand how schema code works is to write schema code. To demonstrate how schema works, this section will demonstrate writing schema in the two most common formats: JSON-LD and Microdata.

JSON-LD Example: Organization Schema

The Organization schema type structures information about an organization. The code for this schema type is commonly placed on the website's home page and can be used by Google to add data to knowledge panels. Currently, Google requires two properties:

- Website URL (in the url property)
- Company logo (in the logo property)

Along with the required properties, it is common to provide the organization name (in the name property), the street address (in the address property), additional links to other websites about this organization (in the sameAs property), and information about the people involved (either in the employee or founder properties).

To begin writing this in JSON-LD, first open the <script> tag that will contain the JSON-LD code. Then, start the JSON-LD object with the opening brace.

```
<script type="application/ld+json">
{
```

The first line declares the @context property as schema.org, and the second line specifies the schema type, in this case Organization, in the @type property. The JSON specification requires a comma be added after each property, which is declared to separate one property from the next.

```
"@context": "http://schema.org",
"@type": "Organization",
```

Once the context and type have been declared, the structured information about this organization can be added to the schema code. Remember, the information stated in the schema needs to also be presented within the page content itself to avoid manual actions. For this example, assume that all information presented is also available on the page.

The first pieces of information to add will be the organization's name, legalName, URL, logo, and email. Each piece of information is stated within a property on a separate line.

```
"name": "Elementive",
"legalName": "Elementive Marketing Solutions",
"url": "https://www.elementive.com",
"logo": "https://www.elementive.com/logo.png",
"email": "info@elementive.com",
```

The next piece of information to present is the organization's address. The properties added so far, like the name or legalName properties, are simple name-value pairs. For example, within the property of "email," the name is "email" and the value is "info@elementive.com". The address property works differently and contains an object, and that object is another schema.org type—that process of containing one schema.org type in another is called nesting.

To begin nesting address, state the name of the address property (address) and then add an opening brace. The opening brace indicates the start of a new JSON object:

```
"address": {
```

On the next line, state what type of object is being added. In this example, the schema.org type is PostalAddress. The following code declares the @type of PostalAddress and then provides the city (addressLocality), state (addressRegion), ZIP Code (postalCode), and street address (streetAddress):

```
"@type": "PostalAddress",
"addressLocality": "Centennial",
"addressRegion": "CO",
"postalCode": "80112",
"streetAddress": "8200 S Quebec St."
```

Once that information has been added, add a closing brace. This brace indicates the end of the address property's object and ends the nesting. Then, add a comma after this brace because the end of this brace indicates the end of the address property:

```
},
```

The final piece of information added within this example will be the names of the people involved in the organization. In this case, the names will be presented in the founder property. The founder property demonstrates another type of data: an array containing two different objects. An array in JSON is noted with a bracket. State the founder property and add an opening bracket to begin the array:

```
"founder": [
```

There are two founders of this organization, so the founder property contains an array of two objects, each containing a Person schema type. Each object begins with an opening brace and ends with a closing brace. The @type of Person is declared for each. Also, note that there is no comma at the end of the founder property because this is the last item presented within the array.

```
{
"@type": "Person",
"name": "Andrea Streff",
"jobTitle": "Consultant"
},
{
"@type": "Person",
"name": "Matthew Edgar",
"jobTitle": "Consultant"
}
```

Once the objects have been added, close the array with a closing bracket:

```
]
```

Lastly, the schema is finished with a closing brace that corresponds to the opening brace on the first line of code. Then, the script tag is closed:

```
}
</script>
```

Once written, the contents of the <script> tag can be placed anywhere on the page. Schema markup written in JSON-LD can be detected in the <head> or <body> of an HTML document.

Microdata Example: Simple Review Schema

Reviews and ratings can be marked up in schema code, and that markup can be used by Google to enhance search result listings. The terms "rating" and "review" are often used interchangeably, but it is important to differentiate between the two within schema. The term "review" means the content discussing the item being reviewed, while the term "rating" refers to the numerical score assigned.

There are two schema types supported by Google for reviews and ratings:

- **Simple review**: This is an editorial review written about something—like a review written about a product by a magazine. The simple review contains a numerical score, but that score is only from one source. The simple review often includes review text as well.

- **Aggregate rating**: This is an average rating of something—like an average rating of a product on an ecommerce website. The aggregate rating contains a numerical score, and that score is an average across multiple sources. The aggregate rating can also contain review content provided by those multiple sources.

This example will demonstrate how to mark up an editorial review in the simple review schema using the Microdata format.

Where JSON-LD is separated from the HTML, Microdata is added as attributes within the HTML tags. For this example, here is the starting state of the HTML code for the review that will be marked up:

```
<body>
  <div>
    <h1>Review of the <strong>Green Widget App on Android</strong></h1>
    <h2>By Jane Smith at Acme Review Co.</h2>
    <p>December 1, 2022</p>
    <p>4 out of 10 stars</p>
    <p>This is the text of the review.</p>
  </div>
</body>
```

When working with Microdata, it is important to remember that HTML is a hierarchy. Anything assigned to a higher level will be assigned to a lower level. The preceding HTML example can be represented hierarchically as

- body
  - div
  - h1
  - strong
  - h2
  - p
  - p
  - p

Microdata relies on this hierarchy to communicate the schema's structure. To communicate this structure, Microdata adds three attributes to the HTML tags:

- **itemscope**: This attribute declares that a new schema item, or scope, is being created. The scope is assigned to a particular HTML tag. Any additional tags under a tag with an itemscope attribute in the HTML hierarchy will be considered as part of that scope.

- **itemtype**: This attribute defines what type of schema is being declared. The value of this attribute is set to the URL for the appropriate schema type.

- **itemprop**: This attribute states the name of the schema property.

The first step for marking up the example HTML code is declaring that this code contains a review. This is done by adding an itemscope attribute to the tag that contains the review—in this case, the <div> tag. With that declared, all tags under the div tag in the HTML hierarchy are part of this Review schema scope. The itemtype attribute is then added and set to the URL representing Review schema.

```
<div itemscope itemtype="https://schema.org/Review">
```

The next step is defining each individual piece of information within this review, starting with the name of the item reviewed which is defined in the itemReviewed property. In this example, the item's name is contained within a <strong> tag. There is nothing special about the <strong> tag—this could just as easily be a <span>, <em>, <p>, or any other tag that can contain other HTML tags.

The itemReviewed property does present a challenge, because it contains two distinct pieces of information about the item reviewed: the application's name and the application's operating system. To communicate this complexity within the schema, the itemReviewed property needs to be defined as a different type of schema. In this case, the item being reviewed is an application and needs to be defined within the schema type of SoftwareApplication.

In order to define the itemReviewed as a type of SoftwareApplication, three attributes need to be added to the <strong> tag. The first attribute added is the itemprop attribute which names what property the <strong> contains and is set to the value of itemReviewed. The second attribute to add is itemscope, to declare a new schema scope, and the final attribute to add is itemtype, which states that this scope defined a software application.

```
<h1>Review of the <strong itemprop="itemReviewed" itemscope itemtype="https://
schema.org/SoftwareApplication">
```

Now, any content within the <strong> tag will be contained within the scope of the SoftwareApplication schema. Because the <strong> tag is contained with the <div> tag, the SoftwareApplication scope will be nested under the Review schema scope.

Next, the individual pieces of information need to be defined, starting with the application's name in the property of "name." When using Microdata, each piece of information needs to be contained within a separate HTML tag. So far, existing HTML tags have been used. However, the text of the name "Green Widget App" is not contained in a separate HTML tag. It is contained within the <strong> tag, which also contains a separate piece of information (the name of the operating system). As a result, a new HTML tag will need to be added around the text of the name to separate it from the name of the operating system. The easiest tag to add, and one that will not add any additional formatting, is the <span> tag.

After wrapping the name of the application in a <span> tag, add the itemprop attribute to the <span> tag and set the value of the itemprop attribute to name—this defines the text within this <span> tag as the name of the application being reviewed.

```
<span itemprop="name">Green Widget App</span>
```

The same process applies to the next piece of information: the name of the operating system. Here again, a <span> tag needs to be added around this text to separate this piece of information. Once the <span> tag is added, the itemprop attribute needs to be added with the value set to operatingSystem—this defines the text within this <span> tag as the name of the operating system.

```
on <span itemprop="operatingSystem">Android</span>
```

Google currently requires that a SoftwareApplication schema type also contain other pieces of information, including the application category. The application category, however, is not contained within the original HTML text. As a result, it needs to be added, and it needs to be added within the <strong> tag because the <strong> tag contains the SoftwareApplication scope.

The simplest method is adding the text to the page in this location and marking up that text similar to the operating system or application name. However, the text cannot always be added to the page in this specific location, as that information fits better elsewhere on the page. In those situations, a common approach within Microdata is adding a <meta> tag within the HTML code that contains the required piece of information. That way, the <meta> tag allows the information to be stated within the schema even though the information is contained somewhere else in the page's HTML.

Here is an example of how this <meta> tag would look when added. The <meta> tag has the attribute itemprop set to the name of the property, in this case applicationCategory, and the <meta> tag's content is defined within the content attribute. For applicationCategory, there are only a handful of supported values—in this case, the most appropriate value is BusinessApplication.

```
<meta itemprop="applicationCategory" content="BusinessApplication">
```

After adding the <meta> tag, all required properties have been added for the SoftwareApplication schema type. As this schema type was defined within the <strong> tag, now that tag can be closed—along with the <h1> tag that contained the <strong> tag.

```
</strong></h1>
```

The next pieces of information to add are the author and publisher of this review. While author is required, publisher is an optional property. These pieces of information are both contained within the <h2> tag. Similar to the itemReviewed property, author and publisher are more complex types of information, and this information must be defined with their own scope using distinct schema types. Author uses the schema type of Person, and publisher uses the schema type of Organization. Using the same process discussed earlier for defining SoftwareApplication, the resulting marked up code is as follows:

```
<h2>By <span itemprop="author" itemscope itemtype="https://schema.org/
Person"><span itemprop="name">Jane Smith</span></span> at <span
itemprop="publisher" itemscope itemtype="https://schema.org/
Organization">Example Review Website</span></h2>
```

After author and publisher, the HTML code next presents the date the review was published. Like the publisher property, date is an optional property but can be included to provide more details about the review. In this example, the date presented on the page is not in the correct format. Instead, schema requires dates to be presented in the ISO 8601 format, which uses the format YYYY-MM-DD. This can be added in a <meta> tag directly after the date. The <meta> tag has the itemprop attribute added set to the value of datePublished, defining the name of this property. The <meta> tag's content attribute is set to the date—and matches the date within the <p> tag.

```
<p>December 1, 2022</p>
<meta itemprop="datePublished" content="2022-12-01">
```

Continuing through the HTML code, the next piece of information is the rating for this review, which will be contained in the reviewRating property. Like the itemReviewed property discussed earlier, reviewRating is a more complex piece of information and needs to be defined within a unique schema scope using the schema type of Rating.

```
<p itemprop="reviewRating" itemscope itemtype="https://schema.org/Rating">
```

The Rating schema type requires the property of ratingValue. The bestRating and worstRating properties can also be included. If not included, it is assumed the bestRating is 5 and the worstRating is 1. As this example review does not use a traditional 1 to 5 scale, these properties need to be added within the Rating schema. Note that the worstRating was not defined on this part of the page so that information is added within a <meta> tag.

```
<span itemprop="ratingValue">4</span> out of <span itemprop="bestRating">10
</span> stars
<meta itemprop="worstRating" content = "0"/>
</p>
```

The last piece of information is the review body. This is an optional piece of information but can be defined by adding an itemprop attribute to the containing HTML tag, in this case the <p> tag. The itemprop attribute should be set to the value of reviewBody, which is the name of this information within a review.

```
<p itemprop="reviewBody">This is the text of the review.</p>
```

Finally, the HTML contains the closing <div> tag, which ends the review text. Because the opening <div> began the Review schema type's scope, this closing <div> will end that scope.

```
</div>
```

Automating Schema Markup

Structuring data with schema, whether implemented with JSON-LD, Microdata, or RDFa (not demonstrated in this book), can lead to Google and Bing enhancing the appearance of listings in search results. Those enhancements can increase clicks to the website. To maximize this increase in clicks, schema markup is typically automated so that it can be easily added at scale to multiple pages on the website.

When automated, it is important to confirm that appropriate checks are in place to verify the schema code is used correctly. Some of those checks can be programmatic. For example, if Review schema will be added to hundreds of articles on a website, logic should be added to the code that prevents schema from being added to any articles that do not contain a numerical rating value as part of the review. However, manual checks should be added as part of regular monitoring of the website. Along with manually testing pages and checking schema performance in Google Search Console, crawl tools like Screaming Frog can be used to validate any schema code present on the website.

Because of these checks, automating and scaling schema requires a larger investment. That investment only grows because programmatic checks need to be updated as requirements change, and the time spent on monitoring tasks will increase as more schema is added. As a result, automating and scaling to multiple pages should not be the first priority when using schema. It is often best to write new schema manually, adding it only to a smaller batch of test pages. If that schema helps generate more clicks from search results to the batch of test pages, then the schema code has proven itself a beneficial part of the website's SEO performance, making the investment in automating and scaling the schema code more worthwhile.

Measuring and Monitoring Guidelines: Schema Markup

- Review the website to determine if any schema is currently being used. If schema is being used, does that schema enhance any search result listings at present? If not, what modifications could be made to the schema markup so that it would enhance listings?

- If schema code is currently enhancing search result listings, how much traffic do those search results generate compared to search results within schema enhancements? This can be reviewed in Google Search Console.

- Is the schema used accurately and reflecting the content on the page? All the information presented in the schema should also be viewable within the page content. If information is only provided in the schema code, this can be seen as a manipulative tactic and lead to a manual action.

- Are all required properties used within the schema code? The required properties do occasionally change, so this should be reviewed on a regular basis and adjusted accordingly. As well, check if there are optional properties that could be used to further describe the page's information and enhance search result listing.

- Check that the schema code used is valid by using the validator tool located at validator.schema.org.

Sitemaps

This chapter reviews XML and HTML sitemaps. Sitemaps provide a list of URLs, and that list can help search engine robots discover new content on a website. Along with listing HTML pages, XML sitemaps can also list other files, like images or videos. However, it is important that XML sitemaps are carefully maintained and only list the pages on the website that robots should rank in search results.

What Is a Sitemap?

A sitemap is an organized list of the pages contained on a website. For SEO purposes, there are two types of sitemaps: HTML and XML.

What Are HTML Sitemaps and Are They Needed?

An HTML sitemap matches the website's overall look and feel (it is called an HTML sitemap as it is written using HTML like all other pages on a website). For smaller websites, the HTML sitemap may list every page, but on a larger website, the sitemap may only list the most important pages, such as the primary categories or top-level pages.

HTML sitemaps are optional but can sometimes help robot (and human) visitors find pages on a website. However, if pages are linked clearly and effectively within the website's navigation and content, an HTML sitemap should not be necessary to locate pages.

© Matthew Edgar 2023
M. Edgar, *Tech SEO Guide*, https://doi.org/10.1007/978-1-4842-9054-5_5

What Is an XML Sitemap?

An XML sitemap is written in XML (eXtensible Markup Language) and can help robots locate all pages contained on the website. In contrast, an HTML sitemap can be viewed and used by both human and robot visitors. All pages that robots should potentially crawl, index, and rank should be listed in the XML sitemap. Along with listing a website's pages, an XML sitemap can also list images and videos.

It is important to remember that including a page in an XML sitemap provides no guarantee robots will crawl, index, or rank that page.

XML Sitemap Size Limits

Each individual XML sitemap is limited to 50,000 URLs and can be no larger than 50MB.

What Is an XML Sitemap Index File?

Websites can use multiple XML sitemaps to list the website's pages. For larger websites, this is essential to avoid size constraints. For all websites, however, having multiple XML sitemaps—such as an XML sitemap for each section or page type—can make managing and monitoring the website's content easier. This is discussed more in the "In Practice" section.

When multiple sitemaps are used, all sitemaps should be listed in an XML sitemap index. Example of XML sitemap index:

```
<?xml version="1.0" encoding="UTF-8"?>
<sitemapindex xmlns="http://www.sitemaps.org/schemas/sitemap/0.9">
      <sitemap>
              <loc>https://site.com/main-content.xml</loc>
      </sitemap>
      <sitemap>
              <loc>https://site.com/blog-content.xml</loc>
      </sitemap>
      <sitemap>
              <loc>https://site.com/product-content.xml</loc>
      </sitemap>
</sitemapindex>
```

What Pages Should Be Included in the XML Sitemap?

To make XML sitemaps as useful as possible, the pages listed in the XML sitemap need to be valid pages that a search engine could potentially index. XML sitemaps should not list URLs to pages that

- Redirect somewhere else

- Return an error message

- Are duplicates (or near duplicates) of other pages

- Contain low-quality or thin content

- Contain noindex directives

- Are blocked by disallow statements on the robots.txt file

- Are password protected or require authentication to access

- Should otherwise not be crawled or indexed

As well, only the canonical URL for each page should be listed. See the "In Practice" section for more details about finding these types of URLs in an XML sitemap.

XML Sitemap Structure

The XML sitemap contains the information about each page in a <url> tag. Within the <url> tag, the page's URL can be defined in a <loc> tag. The page's last modification date, the change frequency, and the page's priority can also be defined; however, only the page's URL is required. Last modification, change frequency, and page priority values are suggestions to robots. Google will ignore priority and change frequency information on an XML sitemap but will use the last modification date if it can be verified. Bing recommends including the last modification date.

Example XML sitemap listing the website's home page, with the optional fields presented:

```
<?xml version="1.0" encoding="UTF-8"?>
<urlset xmlns="http://www.sitemaps.org/schemas/sitemap/0.9">
        <url>
                <loc>https://site.com/</loc>
                <lastmod>2022-06-01</lastmod>
                <changefreq>daily</changefreq>
                <priority>0.8</priority>
        </url>
</urlset>
```

Image XML Sitemap

An XML sitemap can also provide a list of images contained on a particular page. The details about images on a page are added to the XML sitemap within that page's <url> tag. Example of image entry:

```
<url>
        <loc>https://site.com/</loc>
        <image:image>
                <image:loc>https://site.com/pic.jpg</image:loc>
        </image:image>
</url>
```

Video XML Sitemap

Any videos that are contained on a page can be listed in that page's <url> entry in the XML sitemap. The following information is required for each video listed:

- Title

- Description

- Thumbnail image

- Video URL—either provided as the direct link to the video file itself or a link to the video player

There are also optional fields that can be provided, including the video's duration, rating, view count, age restrictions, and more.

Example of video entry showing the required tags:

```
<url>
        <loc>https://site.com</loc>
        <video:video>
                <video:title>Video Title</video:title>
                <video:description>About the video...</video:description>
                <video:thumbnail_loc>https://site.com/thumb.jpg
                </video:thumbnail_loc>
                <video:content_loc>https://site.com/vid.mp4
                </video:content_loc>
        </video:video>
</url>
```

Multilingual and International XML Sitemap

If a website's pages are available in other languages, those can be specified inside an XML sitemap. Each page has a unique entry, in a unique <url> tag. Within that <url> tag, the alternate URLs are provided, and an hreflang attribute is added specifying the language in the ISO 639-1 format.

For example, if a page is written in English and has an alternate version written in French, the sitemap would list both pages as a unique entry and show the alternate URLs. In this example, the entry would be

```
<url>
        <loc>https://site.com/FrenchPg</loc>
        <xhtml:link rel="alternate" hreflang="fr" href="https://site.com/
        FrenchPg" />
        <xhtml:link rel="alternate" hreflang="en" href="https://site.com/
        EnglishPg" />
</url>

<url>
        <loc>https://site.com/EnglighPg</loc>
        <xhtml:link rel="alternate" hreflang="fr" href="https://site.com/
        FrenchPg" />
        <xhtml:link rel="alternate" hreflang="en" href="https://site.com/
        EnglishPg" />
</url>
```

A website's pages target a different country, those can also be specified in the XML sitemap. The entry is similar to the multiple-language entry shown in the preceding example except the hreflang value also includes a country specification in the ISO 3166-1 format. For example, there may be a page in English for visitors in the United States with American spellings and another page in English for visitors in the UK with British spellings. In that example, the XML sitemap entries would be

```
<url>
        <loc>https://site.com/USA-Page</loc>
        <xhtml:link rel="alternate" hreflang="en-us" href="https://site.com/
        USA-Page" />
        <xhtml:link rel="alternate" hreflang="en-gb" href="https://site.com/
        UK-Page" />
</url>
```

```
<url>
        <loc>https://site.com/UK-Page</loc>
        <xhtml:link rel="alternate" hreflang="en-us" href="https://site.com/
        USA-Page" />
        <xhtml:link rel="alternate" hreflang="en-gb" href="https://site.com/
        UK-Page" />
</url>
```

Additional XML Namespace (XMLNS) Attributes

XML sitemaps with additional types of content require specifying an additional XML namespace in an XMLNS attribute in the <urlset> tag.

Image XMLNS

```
xmlns:image="http://www.google.com/schemas/sitemap-image/1.1"
```

Video XMLNS

```
xmlns:video="http://www.google.com/schemas/sitemap-video/1.1"
```

Multilingual and International XMLNS

```
xmlns:xhtml="http://www.w3.org/1999/xhtml"
```

Submitting XML Sitemaps

XML sitemaps can be submitted to Google or Bing directly via Google Search Console and Bing Webmaster Tools. Once submitted, both platforms will report on any errors that were detected within the URLs submitted inside the XML sitemap file.

It is also customary (though not required) to list the URL to the website's XML sitemap (or the URL to the XML sitemap index file if there is one) on the robots.txt file. Example:

```
Sitemap: https://site.com/sitemap.xml
```

In Practice: Reviewing and Cleaning XML Sitemap Files

XML sitemap files, when correctly created, will only list the website pages that should rank in search results—pages that return status 200, are indexable and crawlable, use the canonical domain, do not contain errors, and are free of content-related issues. Whether manually written or automatically generated, XML sitemap files can unintentionally list pages that do not match these conditions, resulting in the XML sitemap listing pages that robots should not crawl, index, or rank. Alternatively, but equally problematic, XML sitemaps may not list every page from the website that should rank in search results, creating gaps in how a robot understands the website.

To avoid these problems and ensure the XML sitemap contains an accurate list of all URLs for all the pages that should rank, it is important to thoroughly review and clean the XML sitemap files using the steps discussed as follows.

Step 1: Validate the XML Document

Robots will only crawl XML sitemap files that are well formed, meaning the XML sitemap file uses the correct XML syntax. The details of XML syntax are outside the scope of this book, but let's review two examples of common errors that can occur.

One syntax requirement in an XML sitemap document is that all elements must have closing tags. Here is an example of an invalid XML sitemap file because the <loc> tag is not closed. Without the closing </loc> tags, robots would be unable to process the XML sitemap document.

```
...
<url>
        <loc>https://site.com/
</url>
...
```

Another common issue is invalid characters used within a tag. For example, the <video:title> tag might use an ampersand character (&), which is an invalid character in an XML document. To avoid this, any text containing special characters can be contained within a CDATA section. CDATA stands for Character Data, and any text contained within CDATA will allow invalid characters because this test will not be treated as part of the XML document. Here is the <video:title> contained within a CDATA section:

```
...
<video:title><![CDATA[My Video Title & More]]></video:title>
...
```

There are a number of free validators available that can be used to validate the XML sitemap file and check for any syntax errors. All syntax errors should be fixed.

Step 2: Check Canonical Domain Usage

The XML sitemap should only contain URLs that utilize the website's canonical domain. As part of cleaning the XML sitemap, the URLs listed should be checked for any noncanonical domain usage. For example, if the canonical domain is https://www.site.com, then the XML sitemap should contain no references to http://www.site.com, https://site.com, or http://site.com. See Chapter 2 for details about selecting a canonical domain.

Not all URLs listed on an XML sitemap will use the same domain. Multilingual websites might be hosted on a subdomain or a separate domain that uses an appropriate country code. As well, images and videos might be hosted on a subdomain, such as images.site.com, or may be hosted on a separate domain for a content delivery network (CDN). This is acceptable, but each of these domains or subdomains should also represent the canonical version of that subdomain or domain—if https://images.site.com is the canonical version of the image subdomain, then http://images.site.com should not be listed on the XML sitemap.

Be sure to correct any noncanonical domains used on the XML sitemap. If the incorrect canonical domain is automatically generated, it is also important to correct the underlying code that generates the XML sitemap file to prevent this issue in the future.

Step 3: Check for Errors, Nonindexable, or Noncrawlable Pages

The XML sitemap should only list pages that are indexable, crawlable, and do not contain errors. Therefore, each URL should be crawled in a crawl tool. For example, in Screaming Frog this can be done by changing to List mode and then uploading the XML sitemap.

Once the crawl has been run, identify any URLs that return the following:

- **301, 302, 307, or 308 response status code**: These response status codes indicate the URL redirects somewhere else (see Chapter 8 for details). Along with removing redirected URLs from the XML sitemap file, it is also important to check if the redirect destination is listed in the XML sitemap file (assuming it should be listed). In some content management systems, the URL might change and a redirect might be added to take visitors to the new URL, but the old URL is still listed in the XML sitemap file while the new URL is not.

- **404 or 410 response status code**: These response status codes indicate the URL could not be found on the server (see Chapter 7 for details). Robots will not index these URLs, so any 404 or 410 URLs should be removed from the XML sitemap file.

  Soft 404s should not be listed in the XML sitemap either. However, crawl tools typically cannot detect Soft 404s by default. If Soft 404s are a known issue, the custom extraction feature can be used within crawl tools to detect a Soft 404. For example, if the Soft 404 page states the error message "this is an error" in an h2, then a custom extraction could retrieve the text of all h2 tags during the crawl. Then, after the crawl, that extracted text could be reviewed to identify any URLs containing h2 tags with the text "this is an error."

- **403 response status code**: This status code indicates the URL requires authentication. Because robots will be unable to authenticate by logging in to a system, and because pages that require authentication are generally poor landing pages from organic search results, any pages that require authentication should be removed from the XML sitemap.

- **5xx response status code**: Any URL returning a response status code in the 5xx category contains a server error. In some cases, these server errors are false positives resulting from problems with how the crawl tool connected with the website. Once false positives are removed, any URLs returning a 5xx response status code should be removed from the sitemap.

- **Noindex URLs**: A meta robots noindex tag (see Chapter 1) indicates a specific URL should not be indexed, while that URL's inclusion on the XML sitemap suggests the opposite. As a result, listing URLs with a noindex tag on the XML sitemap creates conflicting signals that can occasionally lead robots to ignore the noindex tag. To avoid conflicting signals and any resulting problems, it is best to check if any URLs listed in the XML sitemap contain a meta robots noindex tag (for images or videos, an X-Robots noindex can be used instead of the meta robots). If any URLs containing a noindex are found, those should be removed from the XML sitemap.

However, some content management systems do not allow URLs with a noindex to be removed from the automatically generated XML sitemap file. If the code cannot be adjusted, the noindex URLs will have to be left in place. In those cases, it is important to carefully monitor the URLs containing a noindex to ensure robots are respecting the noindex and not including the URL in search results.

- **Disallow URLs**: Robots will attempt to crawl any URLs listed in the XML sitemap. Because of this, the XML sitemap should only contain URLs that robots can crawl. If a URL has been blocked from crawling via a disallow statement on the robots.txt file (see Chapter 1), that URL should be removed from the XML sitemap file. As with the noindex tag, some content management systems cannot automatically remove disallowed URLs, and those URLs will have to be left in place on the XML sitemap file.

Step 4: Check for Content-Related Issues and Missing Pages

Any pages containing content-related issues should also not be listed in the XML sitemap. These content-related issues were discussed in greater detail in Chapter 3, including discussing ways of detecting those issues with a content assessment. Any URLs containing content-related issues identified in the content assessment should not be included in the XML sitemap.

Along with using the content assessment to determine what URLs to remove due to content-related issues, the content assessment can also be used to detect any URLs that are not currently listed in the XML sitemap, but should be. It is important to understand why certain URLs were not listed in the XML sitemap as this could highlight a problem within the code that automatically generates the XML sitemap. Fixing that problem would ensure all future URLs are properly listed in the XML sitemap.

However, URLs missing from the XML sitemap do not always indicate a technical problem. It could be the content management system is incapable of listing non-HTML documents, like PDFs, in the XML sitemap. In those cases, there will always be some number of URLs not listed in the XML sitemap, and work will need to be done to expose those URLs to robots in other ways, like additional internal links.

Step 5: Check Videos and Images Are Listed Correctly

The next step is checking any images or videos listed on the XML sitemap. The first part of checking images or videos was already discussed in Step 3—the images and videos should be crawlable, indexable, and should not return response status codes indicating an error. Along with those checks, images and videos also need to be checked to confirm that the file is contained on the indicated page. Depending on how the images or videos are added to the XML sitemap, there is a possibility that an image will be removed from the page before the XML sitemap entry is removed.

In this example, if pic.jpg is not contained on the website's home page, that would make this entry inaccurate:

```
<url>
        <loc>https://site.com/</loc>
        <image:image>
                <image:loc>https://site.com/pic.jpg</image:loc>
        </image:image>
</url>
```

There is no easy way to check image or video inclusion on a page in bulk. Most crawl tools will miss images that are loaded via JavaScript (including images that are lazy loaded). As well, crawl tools may not detect all videos on the page depending on how the video is embedded. There are custom solutions that can help check for video or image inclusion—for example, custom code can be written to query the database entry for a given page and then check if the image is included in the page's HTML. For larger websites using many images, building this type of custom solution to check for image or video usage can make sense to prevent issues with the XML sitemap.

An alternative approach, though less thorough, is to spot-check several image or video entries on the XML sitemap. If all images and videos are found on the page after spot-checking a sample of 10–15% of images and videos listed, then it is likely no sitewide issues are present. To make this more effective, the sample pages that are checked should be drawn from all the templates and page types across the website—for example, there may be no issues present for images or videos listed for blog posts, but there are issues present for images or videos listed for product pages.

Step 6: Check Multilingual and Internal URL Listings

Any multilingual or international URLs presented in the XML sitemap also need to be reviewed. Along with the checks in Step 3, the review needs to check that the relationship between URLs is correctly communicated and check that the hreflang attributes for language and country are correctly specified.

For example, this entry in the XML sitemap indicates that `https://site.com/EnglishPg` is also available in Spanish and French, indicating each page contains the same information just translated into a different language. However, if `https://site.com/FrenchPg` is not a translation of the same information but instead presents different information than the English or Spanish versions of the page, then the relationship between these pages is not correctly communicated.

```
<url>
        <loc>https://site.com/EnglishPg</loc>
        <xhtml:link rel="alternate" hreflang="fr" href="https://site.com/
        FrenchPg" />
        <xhtml:link rel="alternate" hreflang="en" href="https://site.com/
        EnglishPg" />
        <xhtml:link rel="alternate" hreflang="sp" href="https://site.com/
        SpanishPg" />

</url>
```

The preceding example also contains another error: the Spanish page's hreflang attribute should be "es" instead of "sp." With the language incorrectly specified, robots would be unable to understand what language the page targets.

As with images or videos, these types of issues can be difficult to check in bulk. While crawl tools might be able to detect invalid hreflang specifications, crawl tools would be unable to determine if each version presents the same information, just translated. Instead, the best approach is to spot-check a random sample of 10–15% of alternative version entries to determine if there are any issues.

Step 7: Check Google Search Console and Bing Webmaster Tools

Google and Bing will report on issues their robots have detected with an XML sitemap in Google Search Console and Bing Webmaster Tools. If no issues have been found, the reports will indicate the XML sitemap has been read successfully.

In Google Search Console, errors can be found by clicking a particular XML sitemap and clicking Index Coverage. This will show which URLs on the XML sitemap have or have not been indexed. This is why having an individual XML sitemap for each section of the website can be helpful—with individual XML sitemaps for each section, the issues shown in the report can be isolated to one part of the website. For any URLs listed in the XML sitemap that are not indexed, the report will display which issues are present for those URLs.

In Bing Webmaster Tools, the XML sitemaps can be filtered to only show files with warnings or errors. From there, clicking the XML sitemap's detail page will show which errors or warnings have been detected by Bingbot.

It is important to note that an XML sitemap can be labeled as successfully processed in Google Search Console and Bing Webmaster Tools even if there are issues present with some of the URLs listed. For example, if URLs with content-related issues are listed, that would not prevent robots from successfully processing the XML sitemap file. As a result, while it is helpful and important to review these reports in Google Search Console and Bing Webmaster Tools, they should not be relied upon as the sole method of checking for issues within a website's XML sitemap.

XML Sitemap Checking and Cleaning Schedule

The steps discussed earlier are not simply about identifying a list of URLs to remove, update, or add to the XML sitemap. Instead, the preceding steps are about surfacing deeper issues that are resulting in the problems present on the XML sitemap. For example, a bug in the code might accidentally add URLs containing a noindex to the XML sitemap, but by fixing that bug, that issue will no longer be present on the XML sitemap file. Similarly, the website's architecture might result in PDFs not being listed, but by adding a work-around to the code, the URLs for those PDFs will be listed moving forward.

If those underlying issues are resolved, the XML sitemap will not need to be checked very often because the code generating the XML sitemap can be trusted. In these cases, checking for additional issues quarterly or semiannually for active websites and annually for less active websites is typically sufficient. Along with regular checks, the XML sitemap should be thoroughly tested within any major code updates, like redevelopment projects.

However, it is not always possible to address the underlying issues, and the problematic pages listed (or not listed) will need to be addressed one by one instead. In those cases, the XML sitemap needs to be checked and cleaned more regularly to remove problematic URLs from the XML sitemap—weekly or monthly for more active websites and monthly or quarterly for less active websites.

The process of checking and cleaning XML sitemaps can be overwhelming, especially for large and active websites where there are too many changes to the pages or the code that generates the XML sitemap. As a result, a perfect XML sitemap is not always attainable with time and budget constraints. The goal should not be perfection, but instead to remove as many of the biggest problems as possible and keep the list of URLs in the XML sitemap as accurate as possible. Search engines will not penalize a website for listing too many problematic URLs in the XML sitemap. At worst, search engines may deprioritize and ignore the XML sitemap if it contains too many issues— which does mean one tool to share URLs with robots will have been lost, at least temporarily, but will not result in catastrophic failure for the website's SEO performance.

Measuring and Monitoring Guidelines: XML Sitemaps

- If multiple XML sitemap files are used on the website, check that a sitemap index file exists and lists the correct (and canonical) URL for each of the XML sitemaps.

- Check that all of the website's XML sitemap files are submitted to Google Search Console and Bing Webmaster Tools.

- Regularly review any errors related to the XML sitemaps that are listed in Google Search Console or Bing Webmaster Tools. Address those problems and resubmit the sitemap.

- Review the pages contained in the XML sitemap to identify any pages that should not be included, such as error pages or redirected URLs. See the "In Practice" section for more details.

- Check the website's log files to confirm robots are crawling the XML sitemap(s). This should be checked at least a few times throughout the year or more often for active websites. If robots are not crawling the XML sitemap, review the XML sitemap to determine what issues are present that could be preventing robots from crawling it.

- In Google Search Console, segment the Pages report by the different sitemaps submitted. This is where using multiple sitemaps can be advantageous. For example, if there are XML sitemaps for each site section, the Pages report could be segmented by a sitemap for a specific section, which would help to identify problems that exist within that section.

Page Experience: Core Web Vitals and More

This chapter discusses Google's page experience factors and how those factors affect rankings. This includes speed and Core Web Vitals, as well as mobile usability and interstitials. Along with helping a website rank better in search results, faster speeds and better mobile websites also improve how robots crawl the website. These gains make optimizing speed and mobile usability an important part of improving a website's SEO performance.

What Is Core Web Vitals?

Google's Core Web Vitals ranking factor is part of Google's Page Experience Algorithm. Core Web Vitals is designed to measure three aspects of a website's user experience:

© Matthew Edgar 2023
M. Edgar, *Tech SEO Guide*, https://doi.org/10.1007/978-1-4842-9054-5_6

- First Input Delay (FID)

- Largest Contentful Paint (LCP)

- Cumulative Layout Shift (CLS)

Core Web Vitals does not rely on a robot's interaction with the website but instead measures these three aspects from real people visiting the website (in Google Chrome). This is referred to as field data. Alternatively, Core Web Vitals can also be measured in PageSpeed Insights with a simulated load of the page (which can be referred to as lab data).

What Is First Input Delay?

First Input Delay (FID) measures how quickly the website becomes usable. To be considered good, a website's FID must be under 100 milliseconds. This does not mean that the website needs to fully load within 100 milliseconds only that when a visitor first attempts to interact with the website, the website needs to respond to that interaction within 100 milliseconds. If the website can respond this quickly, the response will feel nearly instantaneous, creating a better overall experience for the visitor.

What Is Largest Contentful Paint?

Largest Contentful Paint (LCP) measures when the largest element on a page is rendered (displayed) in the browser. To be considered good, the largest element needs to load within the first 2.5 seconds. The largest element on a page typically contains the main content of the page, such as a key image or the main block of text. As a result, the longer a visitor must wait for that element to load, the worse the overall experience.

It is important to note LCP measures only elements that are rendered in the visible portion of the users' browser, or the viewport. This means larger elements below the first scroll for visitors will not be counted toward LCP.

What Is Cumulative Layout Shift?

Cumulative Layout Shift (CLS) measures visual stability. If elements move or shift around the page unexpectedly, that can significantly disrupt the visitor's experience interacting with the website. If elements shift in response to a visitor's interaction with a page, that will not present a CLS problem, provided a user understands the shifting is in response to their interaction.

The CLS score is the product of two metrics: the impact fraction and the distance fraction. The impact fraction represents how much space a shifting element uses on the screen. The distance fraction represents how much that shifting element moves. As a result, CLS will be worse when the shifting element takes up more space on the page or when the shifting element moves a greater distance.

Other Speed Metrics

There are many different metrics to measure a website's speed. Along with the Core Web Vitals metrics of FID, LCP, and CLS, there are additional metrics that are helpful to review to understand a website's load time:

- **Time to First Byte (TTFB)**: TTFB measures how long it takes from when a URL is requested to when the first bit of information is returned from the server. Slow TTFB times are correlated with lower performance, including lower search rankings and lower conversion rates. A slow TTFB indicates problems with the website server or hosting configuration. If TTFB is slow, every other speed metric will be slower, including the Core Web Vitals metrics.

- **Start Render** and **First Contentful Paint (FCP)**: FCP and Start Render are both ways of measuring how long it takes between the URL being requested and when something begins to load into the browser. The difference is Start Render measures the rendering of any element, while First Contentful Paint measures the time to render content (like text or images) to the user. If either of these metrics are slow, that indicates the website is loading too many elements (images, JavaScript, CSS, etc.) to the browser. Like TTFB, a slower Start Render time will slow all other metrics.

- **Total Load or Fully Loaded Time**: This metric measures how long it takes between the request of the URL and when everything has finished loading into the browser. Sometimes, this is referred to as DOM Complete, DOM Content Loaded, or Document Complete.

- **Interaction to Next Paint (INP)**: INP measures how quickly the website responds to a visitor's interactions. This is similar to FID, but, unlike FID, INP measures how long it takes the website to respond to all interactions with a page. The INP value reported is the longest response time seen on a page.

How to Reduce Website's Load Time

There are three broad goals to consider when approaching website speed optimization, including optimizing Core Web Vitals:

- **Reduce the total number of files requested**: Each file requested by the website requires another trip to the server to retrieve that file. Even if each trip to the server only requires a few milliseconds, that can add up to a significant amount of time if hundreds of individual files are requested. As a result, every millisecond counts, and every request that can be eliminated, the better. For example, five CSS files could be consolidated into a single CSS file.

- **Shrink the overall size of the website**: The files requested should be as small as possible. The smaller the file, the faster the load. Compressing and minifying CSS or JavaScript files can reduce the total file size, as can removing any unused code. As well, optimizing images and ensuring images are loaded at the proper dimensions can shrink the overall file size.

- **Load the files as efficiently and effectively as possible**: This goal is to make it as simple as possible for a browser or a robot to load the website's files. Caching dynamic files can improve the load efficiency by reducing how many times data needs to be fetched from a database. JavaScript files can be loaded asynchronously, which prevents the script from blocking the website load, or can be deferred, which allows the browser to wait to execute the file when everything else has loaded.

What Is Mobile Friendliness?

Google also includes mobile usability as part of the Page Experience ranking factors. That means, without correctly supporting mobile users, websites can lose traffic from Google. Those factors include

- **Ensuring that a website's text is readable without zooming**: Typically, that means using a font size of at least 12 pixels for the main text on the page, though font sizes as small as 9 pixels will pass Google's mobile usability tests.

- **Sizing content to the screen so users do not have to scroll horizontally to view what is contained on the page**: This applies only to horizontal scrolling on the entire page. Horizontal scrolling on specific elements, if used intentionally, is permissible.

- **Placing links far enough apart so that the correct one can be easily tapped**: This typically means tap targets, like buttons, should be 48x48 pixels with at least 6–8 pixels between them.

Reviewing Mobile Website Methods

Broadly speaking, there are three methods for building a mobile website: responsive website design, dynamic serving, or using separate domains.

- **Responsive website design**: The most common way of creating a mobile website is with a responsive design. With a responsive design, a website's design code (CSS and JavaScript) determines how to transform the page's layout to fit on larger or smaller screens. With a responsive design, there is only one website (with one set of code and content), but that one website responds to different size screens. This is Google's recommended method for creating a mobile website; however, the next two methods can also rank highly in search results.

- **Dynamic serving**: With dynamic serving, the same URLs are used for mobile and desktop websites (and tablet), but the HTML is different across devices. When a visitor requests a page, the web server detects what device a visitor is using and returns the appropriate code and content. There is a single website and a single URL for each page, like with responsive design, but unlike responsive design, there is different code (and potentially different content) for each device. Note that dynamic serving is often used alongside responsive design.

- **Separate domain**: With a separate domain, both the URLs and HTML code are different between mobile and desktop devices (or tablet). When a visitor requests a page, the web server detects what device is being used and routes the visitor to the appropriate URL. Note that sometimes this type of dedicated site is referred to as an "m-dot website" as many companies place this at an "m" subdomain (as in m.site.com). This method is not very commonly used anymore and is generally not recommended for most websites.

Mobile Equivalency: How to Remove Content on Certain Devices?

As discussed in Chapter 1, Google crawls mobile first. This means the mobile and desktop websites should contain equivalent content. This is easiest to achieve on responsive websites as there is only one set of code and content for the website. Certain items might be hidden from view with a responsive website, such as a sidebar, but the main content generally is visible on both devices.

With dynamic serve or separate domain mobile websites, however, there is separate code and content for each device, which can cause equivalency issues. For example, the home page text might be updated on desktop devices, but due to a technical glitch, that update is not applied to the home page text on mobile devices. In this example, the home page text would differ across devices and could cause Googlebot to miss key information, which could harm rankings and performance.

Interstitials, Modals, Dialog Boxes, and Pop-Ups

Many websites use interstitials to promote key calls to action, like newsletter sign-ups, app downloads, sales alerts, and more. These interstitials are sometimes referred to as pop-ups, modals, or dialog boxes, and while there are differences between these terms, the general idea remains the same: some type of notification is shown to visitors when they first load the website, and that notification completely interrupts what the visitor wanted to do, or visitors loaded the website to interact with the website's content, but instead must close the interstitial first.

If interstitials are shown when loading a website on mobile devices and those interstitials block the visitors' view of the website's main content, Google may suppress mobile rankings. Google's recommendation is to convert the interstitial to something less obtrusive that will not prevent visitors from seeing the website's main text, like banner advertisements embedded within the content instead.

This only applies to mobile websites, not desktop websites. For SEO purposes, it is acceptable to use interstitials when the website is first loaded on desktop (though there may be other UX-related reasons not to do so).

As well, this only applies to interstitials that appear when the website is first loaded. If an interstitial is used elsewhere, like when clicking a link to another page of the website or when attempting to exit a website, there will be no negative SEO impact (though, again, there may be other UX-related reasons not to use these types of interstitials).

Mandatory Interstitials

Many websites are legally required to show interstitials to obtain a visitor's consent before the visitor can access the page's main content. There are a few common types of interstitials. The primary examples are age gates that verify a visitor is over a certain age and cookie consent pop-ups for analytics tracking. Google will not penalize these types of interstitials.

It is important that mandatory interstitials show and process the notification on the requested URL instead of redirecting to a different URL. For example, if a visitor requests the URL site.com/age-restricted-text.html, the age gate interstitial should show and process the verification on the requested URL instead of redirecting to site.com/consent.html to show and process the verification.

As well, and if technically possible, the content should still be loaded into the background of the website behind the mandatory interstitial. Because the content will still be available in the code, Google's robots will be able to crawl and index the page's text and include it in appropriate search results.

Safe Search and Meta Rating

If a web page shows explicit or adult-oriented content, a meta tag with the name attribute set to rating and the content attribute set to adult (or RTA-5042-1996-1400-1577-RTA) should be added to the page's head. This indicates the content presented on this page may not be appropriate for all visitors and that Google should restrict the page from certain search results. Example:

```
<meta name="rating" content="adult" />
```

In Practice: Using Split Tests to Optimize Website Speed

The fastest-loading web page only contains plain text without any scripts, styling, images, videos, or other features. This is not realistic. Any modern website must load a myriad of stylesheets, images, videos, and JavaScript files to produce the various features visitors require. Even if each of those files is perfectly optimized and loads as quickly as possible, the inclusion of these files will result in slower speeds.

Improving a website's speed requires making trade-offs between faster load times and all the other functionality or features that could be added to the website. In some instances, the additional features are more important than improving a website's speed. However, it is important to determine which specific files, producing which specific functionality, are truly necessary and worth the trade-off.

For certain files, determining necessity can be subjective. Scripts loaded for analytics tools provide a good example. There is no objective way of determining if loading these scripts to every page, and slowing down the website as a result, is beneficial. Instead, the company managing the website has to assess the subjective value provided by the analytics tool relative to the slower speeds. The value of measuring the website's performance and building audiences within the analytics tool is typically a worthwhile trade-off. However, some companies choose to use multiple analytics tools causing considerably slower speeds, where one analytics tool might suffice.

The necessity of other files, however, can be determined objectively against the website's performance. Take, for example, animations, images, or videos that are added to the page to improve engagement or conversions. Engagement and conversion rates can be measured, and it can be determined which of these features contributes to an improved engagement or conversion rate. The best way to determine what helps engagement or conversion rates is through a split test where a certain number of visitors see the website with the additional functionality and other visitors do not. For example, if the visitors who see the functionality convert better than visitors who do not, that suggests the functionality helps conversions, making the slower speeds necessary.

Identifying Longer-Loading Files

To identify which files should be evaluated in a split test, begin by reviewing what files must be loaded to the page and determining which of those files contributes the most to slower speeds. The best way to find this is by reviewing a waterfall report of all files loaded to the website. In Chrome's Developer Tools, this can be found under Network view, but waterfall reports are also provided by speed testing tools like GTmetrix or WebPageTest.

Typically, images and JavaScript files, including third-party JavaScript files, will consume the most time during a website's load. When reviewing the files that take the longest to load (the files with the longest bars on a waterfall report), determine which of these files can be tested objectively and which require a subjective review. For any that cannot be tested objectively, like files related to an analytics tool, those should be reviewed internally to understand if there is a valid business case to continue loading those files on the website.

After identifying any longer-loading files that can be tested objectively, it is important to check if there are any obvious speed improvements that could be made to any of these files. If an image is currently 2MB, it can be optimized by reducing the file to a smaller size, thus improving the website's load time. Evaluating the image further in a split test would likely be unnecessary because, once optimized, the image would have a negligible contribution to the website's speed. However, any larger, longer-loading files that can be tested objectively and have already been optimized as much as possible are worth evaluating within a split test to determine if the file is necessary and worth loading.

Determine the File's Purpose

Before evaluating the file in a split test, the next thing to review is what each file does on a given page—or the file's purpose. In some cases, the purpose is straightforward, such as an image. In other cases, figuring out what the file does requires investigation. For example, it might not be immediately obvious what features are output by a particular JavaScript file. Instead, understanding what that JavaScript file does requires loading the website with and without that JavaScript file to see what functionality is no longer present.

After determining each file's purpose, there may be files that are simply too important and cannot be removed regardless of the potential speed improvements. In that case, there is no point in evaluating that file in a split test. An image might be the longest-loading item on the page, but if the image contains important content—like an image of a chart or graph—there is no way the page could deliver a high-quality experience, to robots or humans, without that particular image.

Alternatively, reviewing what each file does might identify files that do nothing on the page. In that case, those files need to be removed completely, not evaluated further in a split test.

It is likely there will be files with a questionable purpose—an image might only be included as a decorative element on a page, or a larger JavaScript file might add animation, but not core functionality. These are ideal files to evaluate in a split test.

Group Files by Functionality

All of the steps discussed so far should result in a list of longer-loading files that are already well optimized, have a questionable purpose, and can be evaluated objectively within a split test. As a last step before beginning the test, this list of files should be grouped by associated functionality. It is best to limit each split test to evaluating a single piece of functionality on the website. For example, a calculator might require eight files to load—two JavaScript files, one CSS file, and five images—and all eight of those files could be evaluated in a single split test, but other longer-loading files not associated with that calculator functionality could be saved for additional split tests.

Structuring the Split Test

The general idea behind a split test is fairly simple: split the visitors to the page into two groups, with each group seeing a slightly different version of the same page. Then measure the resulting conversion or engagement rate for each group.

Group A will not see the output of the longer-loading file(s) and Group B will. This makes Group B the control, loading the website as normal with the files being evaluated. Because this test is to evaluate the impact this file has on conversions or engagement among human visitors, robots should be included in Group B—the impact this file has on robots can be evaluated separately.

There is a confounding variable to consider: if those longer-loading files are removed entirely for Group A, then the website will naturally load faster for visitors in Group A. The faster load times could cause Group A to engage or convert more than visitors in Group B. As a result, it would not be clear how the functionality affected engagement or conversions—did conversions or engagement improve simply because of the faster load times or because of the change in functionality?

To address this, the longer-loading files that are being evaluated should still be loaded for Group A, but the output of those files should not be shown. Returning to the calculator example mentioned earlier, all eight files connected with that calculator should still be loaded for Group A, but the key difference would be that Group A should not see the calculator on the page. As another example, when testing font files, the font file itself should be loaded, but the CSS should not use that font. Structuring the test this way means that only the difference in functionality is evaluated within the split test, not the difference in speed.

Evaluating the Results

Following the split test, there are three likely outcomes: positive, negative, or neutral.

A positive result means the test group—the group that did not see the output of the tested functionality—had higher conversion or engagement rates. This suggests the functionality might be hurting website performance and should be removed.

A negative result means the control group—the group that saw the output of the test functionality—had better conversion or engagement rates than the test group. The slower speeds are then necessary because this functionality is important for the website's performance.

A neutral result means that both groups had the same conversion or engagement rates. The functionality tested, then, has no effect. Removing the functionality would make sense—keeping the functionality slows the website's speed but produces no notable benefit.

It is typically best to repeat the same split test multiple times to account for any other confounding variables, like seasonal trends, externalities, or technical glitches.

Speed Is Never the Primary Goal

Achieving faster load times should never be a primary goal for an SEO team or for a company. Instead, the goal of improving speed should always be subservient to other goals. A primary goal is to increase conversions—to the extent improving speed helps with this, then it makes sense to prioritize improving speed. However, adding more images, videos, animations, or other features to a page might also improve conversions. The only way to determine this is by testing those features. If testing shows the various features do help conversions, the slower speeds resulting from images, videos, animations, or other features should be accepted.

The same is true for SEO performance when considering Core Web Vitals. While Core Web Vitals is important, a more important overall goal for SEO is to earn higher rankings by developing authoritative, trustworthy, and high-quality websites. Doing that often requires adding in features or functionality that slow the website. For example, few people (and search robots) would consider an ecommerce website that did not offer robust filtering functionality to be high quality. That robust filtering functionality will probably slow the website's speed, but it is necessary to help the overall website's SEO performance.

Measuring and Monitoring Guidelines: Page Experience

- Review a website's speed metrics, including Core Web Vitals and the other key speed metrics. What metric is performing the worst? How can that metric be optimized? Metrics like TTFB, FID, and INP suggest fundamental issues with how the website's server is configured, while LCP or Start Render suggest issues with how the files are loaded into the browser. Review the speed metrics with each major code update and make the necessary adjustments.

- Remember that while speed is important, there are other factors to consider that will have a greater influence on search rankings. Google will not rank a fast website if the website also contains low-quality content. Similarly, visitors will not convert just because a website loads fast if the content does not also encourage conversions.

- How is the mobile website created? Is it fully responsive or are there elements that are dynamically served? If dynamically served, is the content equivalent between devices? If responsive, are any parts of the main text hidden from view creating an equivalency problem?

- Review the use of interstitials (pop-ups, modals, dialogs) on the mobile version of the website. Do any interstitials appear immediately after the page is loaded and block the main content of the website? If so, are those interstitials legally required or promotional? If the interstitial is promotional, find another way to present the promotional material without using the interstitial for visitors on mobile devices.

Not-Found Errors

This chapter reviews not-found errors, also called 404 errors. These errors can often hurt a website's SEO performance if not properly addressed, as well as hurting conversion rates on the website. While many not-found errors do hurt performance, other errors can help performance. This is because not-found errors can be used as a means of signaling to search engines that problematic pages have been removed. Because of the complexity involved with not-found errors, and the potential impacts on performance, it is important to regularly find and address any not-found errors present on the website.

What Are Not-Found Errors?

A file a robot or human visitor is attempting to access on a website that cannot be found on that website's server is referred to as a not-found error. It can also be referred to as a 404 error, which is derived from the status code that is commonly returned by the server when a requested file cannot be found.

© Matthew Edgar 2023

M. Edgar, *Tech SEO Guide*, https://doi.org/10.1007/978-1-4842-9054-5_7

Defining Related Terms: "Not Found," "404," "Broken Link," "Missing Page," "Broken Page," "Error Page"

The term "not-found error" is the most accurate way to refer to this type of error because the file or page the robot or human visitor was seeking could not be found on the website. Similarly, "missing" or "broken" or "nonexistent" also represents the nature of the error accurately.

As mentioned earlier, the term "404" refers to the status code returned by the server when the server cannot find the requested file. However, the status code returned does not have to be a "404" as will be discussed later. While it is common to refer to these types of errors as a "404," doing so is not always technically accurate.

The term "broken link" refers to the link that led people and robots to the error message, not to the not-found error message itself.

The not-found error is shown to visitors via a page on the website, so the error is sometimes called a "not-found error page," though that term refers to the page itself and not the error encountered by robots and humans.

Sometimes, this error is referred to as an "error page," though there are many types of errors and many types of pages for those errors, so simply referring to a not-found error as an error page is often too vague.

Why Do Not-Found Errors Matter?

For human visitors, a not-found error page represents a dead end. This creates a poor experience as people came to the website hoping to find a page, image, or some other file but could not. If not-found errors are not appropriately handled and minimized, this poor user experience can reduce the website's engagement and conversion rates.

For robots, a not-found error can waste the website's crawl budget on URLs that do not need to be crawled (see Chapter 1). This can cause robots to spend time crawling through not-found error pages instead of crawling through the working pages on the website. However, not-found errors are also an important means of signaling to robots that certain pages have been removed from a website. For example, if pages containing low-quality content (see Chapter 3) have been removed, then those pages' URLs would now return a 404 response, which would tell robots the low-quality pages are no longer on the website.

What Are Broken Backlinks?

A large part of SEO involves acquiring links from external websites—these links are referred to as backlinks. To greatly oversimplify, the more backlinks referencing a website (especially high-quality backlinks), the better that website's chances are of ranking highly in search results and getting more traffic from the search results.

If any backlinks lead robots or people to a not-found error page, those links become a "broken backlink." When backlinks break, any benefit the backlink had in helping a website rank in search results will vanish. This can cause search rankings to drop and result in a loss of traffic from organic search. By fixing broken backlinks, the benefits can often be restored, which can help websites regain rankings and search traffic.

These external broken backlinks can be located by using backlink analysis tools. These tools have a variety of criteria explaining the potential value of the broken backlink were it to be restored. This data helps prioritize what errors to fix first, which is especially helpful when there are many broken backlinks.

What Are Broken Internal Links?

Links within a website can also lead to not-found error pages. Like with external links, internal links help search robots find the pages on a website and help robots determine where a page should rank in search results. As a result, there are SEO benefits when broken internal links are fixed.

Broken internal links have a more direct and negative impact on a human visitor's experience. Visitors click on links to access specific, desirable content, but if the link is broken, the visitor will be left wanting. Fixing broken internal links can often help increase engagement and conversion rates.

Broken internal links can be found by scanning or crawling through a website using a website crawl tool. Most of these tools detect many other issues along with not-found errors.

What Are Other Sources
of a Not-Found Error?

Along with broken links on external websites and within the website itself, broken links can also happen for many other reasons. Human visitors mistype a link—especially when typing links from offline marketing campaigns. Human visitors might follow an outdated bookmark. Or staff members can accidentally share a broken link in an email newsletter or in a social media post.

Whatever the cause, these types of nonlink-based not-found errors can be found by reviewing a website's log files. Alternatively, tracking code can be added to most analytics platforms to track the not-found error URLs visitors reach.

404 and 410 Response Status Codes

The proper HTTP status code for a not-found error page is either a 404 (to indicate not found or nonexistent) or 410 (to indicate gone or removed). Using a 404 status code is appropriate in almost all cases as it indicates the page or file requested is unavailable on the website.

When a page or file is removed from the website, it is technically correct for the URL of the removed page to return a 410 status code to indicate the page or file in question was taken away from this website on purpose. There is some evidence Google may remove pages with a 410 status more quickly from search results and may crawl pages with a 410 status less frequently. If both status codes are used, the 404 and 410 status codes should be used consistently within a website.

What Is a Soft 404?

A Soft 404 error page in most ways looks like a not-found error page except that the server does not return the appropriate status code of 404 or 410, which would indicate the requested file is not found or gone. Instead, the server returns a 200 status code, which indicates everything is "okay" and the file returned normally (even though it did not).

While most humans will not be able to tell the difference between a Soft 404 and a regular 404, without the proper status code robots may not realize the requested file or page cannot be found. This can lead to search engines indexing pages that are a not-found error, which could lead humans to find these error pages ranking in search results.

What Is a Redirected 404?

When a visitor attempts to access a file or page that cannot be found, the website server should immediately return a 404 or 410 status code. Instead, some servers or code configurations redirect the visitor to another location before returning the 404 or 410 status code.

For example, a visitor attempts to access site.com/SomeURL, which is a URL to a page that cannot be found on this website. Instead of the server returning a 404 or 410 when site.com/SomeURL is requested, the server redirects the visitor to site.com/Error where the 404 (or 410) status is returned.

The error URL (in the preceding example, site.com/SomeURL) looks like a redirected URL instead of a URL that cannot be found. With this configuration, robots might not understand the requested file could not be found and might keep the requested file (/someURL in this example) in search results. Essentially, a Redirected 404 is similar to a Soft 404, and, like a Soft 404, the redirected error configuration should be avoided.

How to Format a Not-Found Error Page

- Return proper status code of 404 or 410 so that robot visitors understand the page is not available. Do not allow Soft 404s or Redirected 404s.

- The text on the not-found error page should explain clearly that the visitor has reached an error. For the best experience, the error should avoid blaming the visitor ("the server could not find what you were looking for" instead of "you typed in the wrong link"). The text may differ for 404 or 410 status code pages with a page returning a 404 status code explaining the requested file could not be found and the page returning a 410 status code explaining how the requested file has been removed.

- The text should help a visitor find another place to go, hopefully a place closely related to what the visitor was seeking. Depending on the nature of the business, the error page should also contain contact information to get help. Offering help on the error page can reduce the error's impact on engagement and conversions.

- The error page's design should match the rest of the website's design, including branding and logos to maintain visual consistency between the not-found error page and the rest of the website for human visitors.

In Practice: Fixing 404 Not-Found Errors

The first step to fixing not-found errors is to find every not-found error on the website using the methods described in the "Measuring and Monitoring Guidelines" section. After all errors are found, the next step is determining the proper way to fix the errors.

Broadly speaking, there are four methods available to fix not-found errors:

- Redirect the not-found error to a working page

- Correct the source link leading to the error

- Restore a removed page

- Ignore the not-found error

Deciding which method is the appropriate solution requires knowing more information about each specific error.

Collect Error Information

Begin by creating a new spreadsheet that lists every error URL that has been detected on the website. As with other spreadsheets created throughout this book, create a column with the error URL's path and also a column with the absolute version of the error URL, containing the protocol and hostname. For websites with subdomains, it is also helpful to list the hostname associated with the error URL. An example of this spreadsheet is provided in Table 7-1.

Table 7-1. Fixing Error URLs—Initial Spreadsheet

Error URL	Hostname	Absolute URL
/march-sal	www.site.com	https://www.site.com/march-sal
/an-old-page	www.site.com	https://www.site.com/an-old-page
/cat	www.site.com	https://www.site.com/cat
/2019/promo	sale.site.com	https://sale.site.com/2019/promo
/blogs/a-great-resource	blog.site.com	https://blog.site.com/blogs/a-great-resource

For each error listed, add a column containing the following information:

- **Status code**: Note the specific status code returned for the error. For most errors, this will be a 404, but there may be 410 status codes used on some errors, and other errors might be Soft 404s or Redirected 404s.

  It might seem odd to include Soft 404s and Redirected 404s on this list. However, while the website configuration should be updated to prevent any Soft 404s or Redirected 404s from ever occurring, any Soft 404s or Redirected 404s are still a not-found error that exists on the website. Because the errors exist and could affect performance, these error types should be fixed using the same methods as a properly configured error.

- **Internal link count**: The next piece of information to collect about the error is the number of internal links that reference this error URL, if there are any. This can be obtained by using a crawl tool, like Screaming Frog that reports on the number of Inlinks per page.

- **External link count**: Next, add in the count of backlinks from external websites that reference this error URL, if there are any. This can be obtained from tools like Semrush, Moz, Ahrefs, Google Search Console, or Bing Webmaster Tools.

- **Pageviews**: Knowing how many visitors are reaching the error URL helps in determining how critical an error is. This is often measured with pageviews of the not-found error page. However, sessions, users, or entrances on the error URL can also be measured along with or instead of pageviews. It is typically best to review pageviews or a similar metric over the last few months as some error URLs may only get a few visitors or are subject to seasonality. Collecting this information about the error requires tracking visits to the not-found error page in an analytics tool, like Google Analytics.

- **Search engine robot crawls**: Along with knowing how many visitors reached an error URL, it is important to know if search engine robots are crawling this error URL. The best way of obtaining this number is by reviewing log files. For details on how to review log files, refer to Step 3 of Chapter 1's "In Practice" section. Reviewing a month's worth of crawling data is typically sufficient—the goal of this review is to determine if search engine robots are still aware of the error URL and still see it as an important URL to crawl.

- **Average ranking position**: Another means of establishing priority is checking if the error URL ranks in search results. The average ranking position can be obtained from Google Search Console, Bing Webmaster Tools, or from rank tracking tools like Semrush, Ahrefs, or Moz. It can also be helpful to bring in search volume or impression data from these tools to better clarify which ranking positions matter most for the website's performance. However, if the error page ranks at all, that suggests a greater importance in addressing the error.

- **Reason for error (if known)**: Finally, it can be helpful to add notes to this spreadsheet explaining why this error exists. For example, if the error was intentionally removed, that might change which method is used to fix the not-found error.

After collecting the information, the resulting spreadsheet should resemble Table 7-2. This information will help determine which method is appropriate to fix each specific error.

Table 7-2. Fixing Error URLs—Spreadsheet with Metrics

Error URL	...	Status Code	Internal Links	External Links	Pageviews (3 months)	Crawls (1 month)	Average Position	Reason for Error
/march-sal		404	0	0	15	0	0	
/an-old-page		Redirected 404	13	24	92	18	8	Changed URL structure in 2022
/cat		Soft 404	5	0	8	0	0	
/2019/promo		410	0	81	79	53	4	Deleted page 12-31-19
/blogs/a-great-resource		410	0	62	51	89	14	Deleted page June 2021 as part of content pruning

Method #1: Redirecting Not-Found URLs

A not-found error's URL can be redirected to a working URL on the same website. Chapter 8 contains more details about redirects, though to summarize, with a redirect, the website server routes people from the error page to a working page on the website. A redirect prevents visitors, including search engine robots, from reaching the not-found error URL.

If there are many external or internal links referencing the error URL, redirecting the error URL is typically the best solution. As well, redirects are typically an appropriate solution when search engine robots regularly crawl an error URL months or years after the URL first returned the error.

Importantly though, there must be an appropriate page to redirect to. The page redirected to should contain content that is closely related to the content visitors were hoping to find when they reached the not-found error. If the page redirected to is not closely related, visitors will be confused, and this redirect will create a poor user experience. Robots may also misunderstand the redirect and not update the index correctly. It is almost always better to let the URL continue returning an error, instead of worsening the website's user experience or causing problems for robots. Selecting the appropriate page will be discussed in more detail in Chapter 8.

Based on the example data shown in Table 7-2, it would likely make sense to redirect the error URL /2019/promo to another page on the website, like the current year's promotion page. This is for three reasons:

1. The 2019 promotion page is continuing to be crawled heavily by search engine robots even though the page was removed years ago. Search engine robots find this URL important to crawl. By redirecting the URL elsewhere, that should help search engines treat the page redirected to as equally important and worth crawling.

2. There are several external websites still linking to this URL even though it is in error. Redirecting /2019/promo to another page on the website would also attribute those backlinks to the URL redirected to, potentially helping the performance of that page.

3. Most importantly, there are lots of visitors accessing this page and getting the error—that will undoubtedly hurt not just SEO performance but the overall performance of the website. Related to this, the error URL ranks on average in position 4 on search results, and that means some people are likely finding the error URL in organic search results. Redirecting the URL will prevent visitors from reaching the error URL, improving the overall user experience.

Method #2: Correcting Source Link

Another way of fixing not-found errors is by correcting any links leading visitors or robots to the error URL. However, this fix only applies to links that are under the control of the SEO team or company managing the website. A broken link on the website itself can be almost always be updated to no longer link to the broken page. This is also true for some types of external broken links, like links contained on that company's social or local profiles. Most external broken links, however, cannot be updated—for example, a broken link contained in a news article about the company likely cannot be updated.

Reviewing the example data shown in Table 7-2, the error URLs /an-old-page and /cat both have internal broken links. These internal links should be updated to link to a different URL instead. Alternatively, the internal links referencing these error URLs may no longer be relevant and could be removed altogether.

Typically, updating links should be done alongside redirecting error URLs. Even if the source link is changed to no longer link to the error URL, robots may keep the error URL in a database and will continue to crawl, index, or rank that error URL. As well, there may be links referencing the error URL that cannot be updated, and those links would continue to signal to robots that the error URL should be crawled.

Returning to the example data in Table 7-2, /an-old-page has external links and is crawled by search engine robots. Plus, that error URL ranks in search results. Therefore, along with updating the internal links that reference /an-old-page, a redirect should be added to reroute robots or visitors who attempt to access /an-old-page to a working URL on the website. None of that is true for the error URL /cat, so /cat would not need to be redirected.

Method #3: Restoring Removed Pages

Sometimes, a page is removed from the website, but people and robots still come to the website looking for the removed pages. Because the page has been purposefully removed, visitors who attempt to access the page will reach a not-found error instead. In these instances, one way to fix that not-found error is by restoring the deleted page. Restoring the removed page would let robots and visitors find the intended page instead of reaching the not-found error message.

Restoring a deleted page is a last-resort solution. Often, the page was removed because the content no longer fit on the website. If that is the case, the error URL cannot be redirected because there is no relevant page to redirect to as the related content has also likely been removed. As well, because the page has been removed, there should not be any internal links referencing that removed page.

Whether restoring the page is a correct solution will depend on why the page was removed. In some cases, the page was simply removed because it was outdated. If there is still a lot of demand for removed content, it may be worth restoring the page, though with changes to bring the page up to date. In Table 7-2's example data, this might be the case for /blogs/a-great-resource—with some updates, the old blog post might still be relevant to visitors and could rank better in search results.

In other cases, the page was removed for a specific reason and cannot be restored. See Method #4.

Method #4: Ignoring the Not-Found Error

Sometimes, the not-found error does not need to be fixed. There may be no relevant content to redirect the error URL to. Or no matter how popular the removed page might be among visitors, there is a strong reason to not restore the removed page. Leaving the error in place will eventually cause robots to stop crawling the page and stop including the page in search results.

For example, the page might have contained low-quality content, and the error message is part of solving a content-related problem on the website. In the preceding example table, /blogs/a-great-resource might have been deleted from the website not because of outdated content, but because the blog contained low-quality content. Removing the page is an important step to fix the website's quality and improve SEO performance (see Chapter 3 for more information about content-related issues).

As another example, the page might have contained information about a product that is no longer sold by the company. The error message signals the product's removal. In these cases, it can be helpful if a custom error message is developed for removed products. This customized version of the not-found error message can explain why the product has been removed and might even suggest alternatives to consider instead. Whether customized or not, though, the error message should remain in place.

Preventing Errors by Predicting Typos

The best type of error is the error that does not happen. For URLs, people will type in directly to the browser (such as URLs included in offline sales or marketing materials). It is best to predict some of the common mistakes people may make when typing in that URL. Then, establish redirects taking visitors from the predicted typo version of the URL to the correct, nontypo version.

In the preceding example table, the error URL /march-sal could have been avoided by adding a preventative redirect for the misspelling of "sale." Similarly, preventative redirects could be added for other common misspellings, like /marc-sale.

Regularly Finding and Fixing Not-Found Errors

Not-found errors do not present an equally severe problem for all websites. Active websites, which regularly change URLs and remove pages, will often have more not-found errors and might suffer performance issues because of those errors. In contrast, websites that change less often may never experience that many not-found errors or the related performance issues. However, broken backlinks and typos can occur, causing not-found errors on even the least active websites.

As a result, every website needs to regularly check for and fix not-found errors. The frequency of these checks should change based on the nature of the website. SEO teams managing active websites should check for not-found errors on a weekly or monthly basis, while less active websites should be checked quarterly or yearly. However, all websites should be checked for not-found errors during any major changes, such as redesigns or redevelopments.

When doing these checks, it is critical to remember to look for not-found errors caused by all sources. Visitors, whether human or robot, can find not-found errors through internal links on the website, via broken backlinks on external websites, or by directly accessing old URLs stored in a search engine robot's database (or in a visitor's bookmarks). It is only by finding not-found errors from all sources that SEO teams will be able to fix and prevent not-found errors from affecting the website's performance.

Measuring and Monitoring Guidelines: Not-Found Errors

- Set up the needed tools to detect not-found errors on the website. Broken internal links can be detected with crawl tools, like Screaming Frog. Broken backlinks can be detected with backlink analysis tools, like Semrush, Moz, or Ahrefs. Some not-found errors will not be detected by these tools, so log files can be reviewed to detect any not-found errors accessed by visitors or robots. Similarly, analytics tools, like Google Analytics, can be configured to track visitors to not-found error pages. Using multiple sources ensures that not-found errors can be identified regardless of the source leading visitors or robots to the error. Google Search Console and Bing Webmaster Tools can also be helpful to identify errors, though ideally errors should be caught prior to search engine robots identifying the problem.

- Review the website's not-found error page to confirm it is properly configured and properly formatted. Are proper response codes returned (404 or 410)? Is the error clearly communicated? Is the error page designed to match the rest of the website? Will the error message help to reduce the negative impact on conversions and website engagement?

- Are there Soft 404s or Redirected 404s present on the website? These are typically harder to detect because the status code will not clearly signal an error. Google Search Console will report on Soft 404s identified on the website. Another common method of detecting Soft 404s is reviewing title tags as a Soft 404 will often have a title tag indicating an error. Detecting redirected 404s requires reviewing the redirect destination, which will be discussed in more detail in Chapter 8.

- Review how robots are crawling the website using log file tools or crawl stats in Google Search Console. Are robots spending an inordinate amount of time crawling error pages? If so, what is causing robots to crawl error pages this much? What changes can be made to help robots find the nonerror parts of the website instead?

Redirects

This chapter covers redirects. Redirects can be used to support changes to URLs and are a means of fixing not-found errors. However, if not properly maintained, redirects can create problems for a website. To achieve optimal SEO performance, it is important to understand the different types of redirects that can be used, as well as understand how to properly configure and maintain all of the redirects on the website.

What Are Redirects?

A redirect sends visitors, human and robot alike, from one URL to another. For example, a visitor accesses site.com/A but is redirected to site.com/B.

Redirects can also send visitors to a different domain. For example, a visitor accesses somesite.com/A but is redirected to othersite.com/Z.

The URL redirected from is called the redirect source or redirect origin. The URL redirected to is referred to as the redirect destination or redirect target.

Why Use Redirects?

When updating a website, whether as part of a major overhaul, consolidation of pages, or routine maintenance, occasionally there may be a need to change the URL of a particular page. Alternatively, a page might also be removed completely during these types of pages. Removing a page or changing a page's URL breaks the old URL and anybody attempting to access the old or removed URL will receive an error message saying the page associated with that old URL can no longer be found.

© Matthew Edgar 2023

M. Edgar, *Tech SEO Guide*, https://doi.org/10.1007/978-1-4842-9054-5_8

For human visitors, this hurts conversions and engagement because people will not be able to access the requested page. When search engine robots come across these errors, the broken URL will be removed from the search index, and people will no longer be able to access this changed or removed page from a search result. This can have a negative impact on organic search performance.

Redirects offer a solution to this problem and prevent visitors from encountering an error. Instead of reaching an error, visitors who attempt to access the change or removed URL are redirected to another page.

Redirects can also be added for various marketing campaigns to create a short version of a URL. For example, a promotional URL such as Site.com/BigSale can be included in ads, but when attempting to access that URL, a visitor is redirected to site.com/events/promotions/big-sale.html.

Does the URL Have to Change?

Even when added properly, redirects can have a negative impact on a website's search performance. It takes time for a search robot to find the redirect and assign the various ranking signals associated with the old URL, the redirect source, to the new URL, the redirect destination. Most in the SEO community suspect that not all signals are fully transferred to the new URL, even with the most successful redirects. Redirects can cause the new URL to rank lower than the old URL, resulting in a drop in organic traffic compared to the performance of the old URL. At a minimum, there will often be a drop in performance during the time search robots are discovering the redirect, adjusting the assignment of various signals, and updating search result rankings.

Before changing a URL, ensure the change is for a good reason. For instance, the main topic of the page substantially changed, and that requires a new URL to more accurately represent the page's new topic. Or the website is moving to a new platform, and there is no choice but to change URLs. Even in these situations, as much as possible, limit the amount of URL changes. It is rarely a good idea to change a URL simply in hopes that the new URL will rank better—many rankings have been lost as a result.

If URLs Must Change...

One of the most common reasons for changing URLs is when moving a website to a new content management platform. If a website only has a few pages, redirecting each page from the old to new URL is relatively simple and a worthwhile practice that can be completed manually.

If there are many pages to transition, this will require adding many redirects. However, this does not always require writing many redirects if the URL change follows a specific pattern. For example, there might be thousands of product pages with changed URLs, but the product page URL changes fit a pattern—the URLs might be changing from the format /product/product-name.html to /buy/product-name/. These pattern-based redirects can be handled with regular expressions. A single regular expression redirect can process the redirect for the thousands of changed URLs.

Sometimes, though, regular expression redirects are not possible because old and new URLs do not fit a unique pattern. In these cases, it is more practical to select only the "best" pages to redirect. This may include redirecting pages with several external links, high traffic volume (especially organic traffic volume), high number of social shares, high conversion rates, or high engagement rates.

Where to Redirect?

If the URL for a page is changed but the page itself is kept on the website, it is easy to determine where to redirect to: redirect the page's old URL to the page's new URL. For example, an old blog post might have been republished at a new URL, and the old blog post URL can redirect to the updated blog post URL.

Redirects can also be added for removed pages. For removed pages, the redirect should take visitors from the URL of the removed page to a page that meets similar expectations as the removed page. For example, an old product page was removed, but there is a new version of the product located at a new URL—the old product page's URL can be redirected to the new product's URL. If there is no new version of the product, a redirect could take visitors to a category page that lists products like the one that was removed.

Avoid redirecting visitors to a page that does not represent what they were expecting as this can be frustrating and confusing. If no relevant page exists to redirect to, allow people to see an error message—that is why errors exist (see Chapter 7). By seeing an error, visitors will know the page they were hoping to access is no longer available instead of being confused after being redirected to a different page that does not meet their expectations.

The redirect destination should represent the canonical URL (see Chapter 2). This adds another signal on the website to help enforce the canonical URL decisions when robots crawl the redirects. If a redirect sends robots to a noncanonical version of the URL, that might confuse robots, causing the incorrect URL to rank in search results.

Avoid Home Page Redirects

The home page is almost always broadly focused, so the chance that the home page will be similar enough to the removed page visitors wanted to find is highly unlikely. As well, Google will tend to treat redirects to the home page as if they are a type of not-found error message, which results in the redirect not retaining any value that supports a website's SEO performance. It is best to find a more relevant page to redirect to instead of redirecting to the home page.

What Are Redirect Chains?

A redirect chain is where URLs redirect from one to the next to the next. Each step in the chain is called a redirect hop. Example of a redirect chain:

```
asite.com/A -> asite.com/B
asite.com/B -> asite.com/C
asite.com/C -> asite.com/D
```

In this example, asite.com/A redirects to asite.com/D in three hops, creating a redirect chain. Note that the URL asite.com/B redirects to asite.com/D in two hops, and asite.com/C redirects to asite.com/D directly (a single hop).

Robots waste resources crawling through redirect chains and may simply stop following the chain after a certain number of hops, meaning the robots may not locate the final page in the redirect chain. To avoid chains, update all redirects to go directly to the redirect destination. In the preceding example, asite.com/A, asite.com/B, and asite.com/C should all redirect directly to asite.com/D in a single hop. If chains cannot be avoided, the number of hops in the chain should be reduced as much as possible.

What Are Redirect Loops?

A redirect chain that circles back onto itself is called a redirect loop. No destination can be arrived at by following these redirects, meaning visitors will be unable to access any pages within the redirect loop. Robots will waste crawl budgets, and human visitors will see an error message in the browser. Example of a redirect loop:

```
asite.com/A -> asite.com/B
asite.com/B -> asite.com/C
asite.com/C -> asite.com/A
```

Server-Side Redirects

A server-side redirect happens on the server, which means no content is sent to the browser for the page being redirected. These redirects can be configured in several different ways. One common method is within server configuration files, such as in the .htaccess file (on Apache) or via a web.config file (also Windows). On Windows, redirects can also be configured via IIS. Redirects can also be configured in content management systems—for example, WordPress has plugins that can manage redirects. Another common place to define redirects is within content delivery networks, like Akamai or Cloudflare.

301 and 302 Response Codes

There are two common HTTP response codes returned by the server to indicate a redirect: 301 and 302.

- **301 status code**: Indicates the redirect is permanent. This is the type of redirect that should be used by default, especially when changing a page's URL or redirecting a removed page's URL to some other URL.

- **302 status code**: Indicates the redirect is temporary. There is no standard for how long "temporary" is. With this status code, the redirected URL may show up in search results.

The 302 status code is the default for most server-side redirects. The 301 status code must be explicitly defined.

307 and 308 Response Codes

The 307 or 308 response status codes are available in HTTP 1.1. A 307 response indicates a temporary redirect, and a 308 indicates a permanent redirect. The 308 code is supported by Google and will be treated like a 301 redirect. As both the 301 and 308 status codes are supported, current 301 redirects do not need to be changed to a different status code.

The 307 redirect is often seen within testing tools, including within Google Chrome's Developer Tools, when testing redirects between nonsecure (http://) and secure (https://) websites. The 307 response is returned in the browser even if the server is not configured to return a 307 response code. This is because browsers will send a 307 redirect to a secure connection from a nonsecure connection using an internal redirect instead of connecting with the server to complete the redirect. Robots do not load nonsecure to secure redirects in the same way as a browser and will connect with the server; the server should return a 301 response for a nonsecure to secure redirect.

Client-Side Redirects

Visitors can also be redirected to the new URL via code that runs in the visitor's browser. The browser is referred to as a client, hence the term client-side redirect. Using a client-side method, when the visitor requests a redirected URL, the server returns content to the browser for this URL. The browser then loads the content for that page into the browser. Included in that content loaded into the browser is code instructing the browser to redirect elsewhere. Upon seeing that instruction, the browser runs (executes) the redirect code and redirects the visitor to another URL.

Client-side methods can be blocked depending on a visitor's browser configurations or plugins added to the browser. As well, search robots may ignore client-side redirect methods. For those reasons, client-side redirects are not as preferable as server-side redirects.

Client-Side Redirect Option 1:
JavaScript Redirect

In JavaScript, there are three options to conduct the redirect. The redirect destination is redirect_to_url in the following example. All methods are generally supported by Googlebot when crawling and indexing the website, though server-side redirect will work more consistently.

```
Option #1: window.location.href = redirect_to_url;
Option #2: window.location.replace(redirect_to_url);
Option #3: window.location.assign(redirect_to_url);
```

Client-Side Redirect Option 2:
Meta Refresh

The other client-side redirect method is a meta refresh. This is configured in a <meta> tag located within the <head> of the HTML document.

If a URL is provided within the meta refresh's content attribute, the meta refresh tag will redirect the visitor to that URL. If no URL is provided, the meta refresh will reload the current page.

By default, the refresh or redirect will occur immediately upon loading the page. The meta refresh can instead include a timer to delay the refresh or redirect. Delayed redirects are interpreted by Googlebot as temporary redirects.

Example of a meta refresh that is set to redirect within two seconds of page load:

```
<meta http-equiv="refresh" content="2;url=https://site.com/A">
```

Redirect Map

A redirect map lists all redirects contained on a website, acting as a reference point for the humans who manage that website. The redirect map lists what URLs are redirected (redirect source), where the redirect goes to (redirect destination), and what method is used for each redirect. It is also helpful to include the date the redirect was added or last updated. If the team managing redirects is larger, it can also be helpful to add a reason why the redirect was added and information on who added the redirect.

An example of a redirect map is shown in Table 8-1.

Table 8-1. Example Redirect Map

Redirect Source	Redirect Destination	Method	Date	Reason
/old-page	/new-page	301	12/1/22	Change URL

Three Benefits (Among Others) of Using a Redirect Map

First, a redirect map can help avoid future redirect chains. As an example, a redirect might have been added two years ago taking visitors from site.com/A to site.com/Z. Now content is changing again, and site.com/Z will redirect to site.com/X. In this situation, site.com/A would need to be updated to redirect to site.com/X to avoid a redirect chain from site.com/A to site.com/Z to site.com/X. A redirect map can make it easier to spot any redirects when the website is changing to avoid chains (or loops).

Second, redirect maps maintain a list of temporary redirects (302) to determine what redirects need to be updated or removed. A temporary redirect is, by definition, temporary and should be removed or changed to redirect elsewhere at some future point. If the redirect should not be removed or changed, then the redirect map has surfaced redirects that need to return a 301 status code instead of a 302 status code.

Third, a redirect map can also help highlight any instances where a redirect routes to something other than a page that returns a status 200, such as a redirect destination that returns a status 404 or status 500. The recommended action is to crawl through the redirected destination URLs regularly to confirm those URLs continue to return a 200 status code.

See the "In Practice" section for more details.

How Long to Keep Redirects?

If the redirect uses a 301 HTTP response status code, the proper technical answer is that the response status code indicates the redirect is permanent, and, as a result, the redirect should be kept forever. This is often not practical as keeping redirects indefinitely can result in substantial management challenges, especially as the website is updated and content changes.

To arrive at a more practical solution, there are two main factors to consider to determine how long each redirect should be kept:

- Does the redirected URL have traffic?
- Does the redirected URL have backlinks?

If people or robots are continuing to access a redirected URL, then that suggests the redirect should be kept. For some websites, it makes sense to maintain redirects for URLs with any amount of traffic, while for other websites, it is only important to maintain redirects that receive traffic above a certain threshold (for instance, keep redirects only if they have more than 100 hits over the course of a month).

Similarly, if other websites continue to link to redirected URLs (or if there are internal links within the website referencing the redirected URL), it often makes sense to maintain the redirect as people or robots might use those links. Here again, for some websites it might make sense to maintain redirects for URLs with any number of backlinks, while for other websites, it may make more sense to maintain redirects only if there are more than a certain number of backlinks referencing the redirected URL.

In Practice: Review and Clean Up Existing Redirects

Older websites can end up with lots of redirects. As these redirects age, it is common for redirects to break due to subsequent changes. Redirects might have been configured ten years ago during a website redevelopment project, but most of those redirects might now result in chains, loops, or dead ends because the destination URLs were changed in another redevelopment project five years later and a content refresh two years after that. The older the website and the more update projects the website has undergone, the bigger the problem can become. Even newer websites can suffer from these problems if the website is actively changing URLs and restructuring content. The only solution is to regularly review and update existing redirects.

Step 1: Build Redirect Map

The first step is building the redirect map as discussed in the main section of the chapter. On many websites, redirects are configured in a single location, such as the .htaccess file on Apache servers, or as part of the content management system, like the plugin Redirection in WordPress. In these cases, it is usually easy to export a full list of redirects and bring those into a spreadsheet, creating the redirect map. (This might seem tricky to do for an .htaccess, but remember that an .htaccess file is a space delimited file, similar to a CSV.)

For other website configurations, redirects are managed in multiple locations. Some redirects might be configured on the server's configuration files (like . htaccess), while other redirects are configured in the content management system, and yet more redirects are configured in the content delivery network (like Cloudflare). Client-side redirects are often configured directly within the code itself.

Configuring redirects in multiple locations can have some advantages, like optimizing server performance or simplifying development. However, configuring redirects in multiple locations can make finding all of the redirects challenging. If redirects are configured in multiple locations, it can be helpful to update the redirect map to note where the redirect is configured.

At the end of this step, there should be a full redirect map listing all redirects on the website. For simplicity, the example redirect map in Table 8-2 shows the redirect source and destination URLs as relative paths, assuming both are located on the same domain. This may not be the case for all websites as redirects can lead to subdomains or external websites. As needed, adjust the redirect map to show absolute URLs or include columns for the source and destination hostnames.

Table 8-2. Redirect Map Containing Redirects from All Sources

Redirect Source	Redirect Destination	Method	Date	Reason	Location
/A	/B	301	6/1/22	Redesign	Redirection
/B	/C	301	6/1/22	Redesign	.htaccess
/D	/E	301	7/1/22		Cloudflare
/F	/B	302	9/1/22	Content update	Line 14 of /inc/cmn.php
/H	/Q	JavaScript	10/1/22		Line 12 of /js/main.js

Step 2: Reviewing Redirect Destinations

The next step is checking each redirect destination, which is the URL contained in the "Redirect Destination" column of the redirect map. A redirect review project should check that each redirect destination returns a response status code of 200.

Obtaining the status code for the redirect destination URLs requires crawling each of the destination URLs. As one example, this can be done in Screaming Frog by changing to List mode and then uploading the list of destination URLs.

Once obtained, add the status code for the redirect destination in a new column of the spreadsheet. Any redirect destination URL that returns a response status code of 200 or 304 is likely operating properly. These can be reviewed further in Steps 3 and 4. However, any redirect destination URLs that do not return a response status of 200 present an immediate issue.

Redirects with a destination resulting in a 404 or 410 status code are a Redirected 404 (discussed in more detail in Chapter 7). These redirects either need to be removed or updated. If the redirect is removed, then the redirect source URL will return a 404 status, making it a correctly configured not-found error instead of being a redirected 404. Alternatively, updating the redirect destination will allow the redirect source to continue operating as a proper redirect.

There is a complicating factor when checking for redirect destinations that return a not-found error: Soft 404s. Soft 404s return a status 200 instead of returning a status code that indicates an error (see Chapter 7 for more details). Soft 404s can typically be detected by checking the content of the title or H1 of the redirect destination page. In other cases, these need to be checked for manually.

Redirects with a destination resulting in a 301, 302, 307, or 308 indicate redirect chains or redirect loops. Redirect chains need to be updated to make it easier for search engine robots to crawl the redirects, and redirect loops need to be corrected so that robots and human visitors will not encounter the error. In the example redirect map, there are two redirect chains. The source URLs /A and /F redirect to /B, while /B redirects to /C. The redirect destinations for /A and /F need to be updated to /C so that /A and /F redirect directly to /C instead of redirecting through a chain.

A 5xx status code, such as a 500 or 503, might be returned as well (see Appendix A for more details). Server errors can sometimes be a false positive—the crawl tool used to check the redirect destinations might have been blocked from accessing the website due to firewall settings, resulting in the server error. If the server error is legitimate, then the redirect needs to be updated accordingly.

Occasionally, the redirect destination requires a login, as indicated by a 401 or 403 (Appendix A has more details). Any redirects that route to a login may be accurate. However, these should be verified to ensure no updates are necessary.

At the end of this step, the redirect map should include the redirect destination's status response code and, in a new column, note any immediate actions to take based on what those status codes indicate. An example of the updated redirect map can be seen in Table 8-3, with the actions to take listed in the "Destination Updates" column.

Table 8-3. Redirect Map—After Reviewing Redirect Destinations

Redirect Source	Redirect Destination	...	Destination Status	Destination Updates
/A	/B		301	Update to /C
/B	/C		200	
/D	/E		200	
/F	/B		200	Update to /C
/H	/Q		200	

Step 3: Reviewing Redirect Sources

The next step is reviewing redirect sources. Collecting data about each of the redirect sources will help determine which redirects are still valuable and helping the website's performance.

- **Search engine robot crawls**: The first metric to review is how often a search engine robot has crawled the redirect source. The best way of obtaining this number is by reviewing log files—this was discussed in detail in Step 3 of Chapter 1's "In Practice" section. For redirect cleanup projects, it is best to review at least three months of crawl data as robots may not crawl redirected URLs very often.

  If a robot regularly crawls a redirect source, such as crawling that redirect source daily, then the redirect needs to be kept and, as needed, updated depending on the findings in Step 1.

  If a robot has not crawled a redirect source at all, the redirect should be removed. See Step 4 for possible exceptions.

It is harder to decide what to do with redirects that a robot has only crawled a few times. Each company and SEO team must decide what crawl threshold is appropriate. For example, over a 90-day period, if there are only ten crawls of a given redirect source, should the redirect be kept? Many companies would remove this redirect, dismissing it as too few crawls. Though, perhaps the redirect source was an important page on the website for many years, meaning the redirect ought to be kept. To help decide on these trickier situations, other metrics described in this step should be checked to determine what action to take.

- **Entrances**: Along with robots, human visitors to the website may also use redirects. Visitor usage of redirects can be tracked in analytics tools by adding tracking parameters to the redirect destination. For example, this can be configured in Google Analytics with UTM parameters. It is best to configure the utm_source as the path of the URL redirected from.

  site.com/old-page -> site.com/new-page?utm_source= old-page&utm_medium=redirect

- This will allow the redirect destination to track as a unique session source in analytics reports. In the example shown earlier, a visitor who entered the website on the URL /old-page would have a session source listed of old-page and a medium of redirect for that session. Measuring activity then would require reviewing each session source.

There are two drawbacks to this type of tracking. First, if this tracking is used on internal links, it would reset the session source in analytics reporting, creating two sessions when there was only one session. For example, a visitor may have entered the website from an ad, but if that visitor clicked an internal link that redirects with tracking parameters, then the visitor would also be reported as having entered the website from a redirect.

The second drawback is that this type of tracking means the redirect destination does not reflect the canonical version of the URL. Instead of robots finding the canonical URL when crawling the redirect, robots will find the URL with the tracking parameters included. As discussed earlier, this can cause robots to rank incorrect URLs in search results (in this case, ranking URLs with the tracking parameters).

Given these drawbacks, it is best to only use tracking parameters on redirect destinations in limited instances. One common instance is redevelopment projects where it is critical to have as much data as possible to decide how to handle existing redirects. Leading up to a redevelopment project, redirect tracking parameters can be added to redirect destinations for a few weeks to monitor usage. After enough data is collected, the tracking parameters can be removed.

Once this data is collected, the evaluation is similar to the evaluation of crawls in the log file: Is there enough usage to justify keeping this redirect? The one difference though is that analytics tools represent visitors to the website, not search engine robots. That means the threshold for keeping the redirect might be much lower—after all, if even one visitor converted after entering the website via a redirect, that might be enough to justify keeping the redirect.

- **External link count**: Next, it is important to know if any external websites link to the redirect source URL. This can be obtained from tools like Semrush, Moz, Ahrefs, Google Search Console, or Bing Webmaster Tools.

  In general, the more websites that link to the redirect source, the more important it is to maintain that redirect. This is especially true for any links that reference redirect source URLs contained on high-quality external websites. If there are no links from external websites to the redirect source URL, then the redirect should likely be removed.

- **Internal link count**: Finally, add the number of internal links that reference the redirect source URL, if any, to the spreadsheet. This number can be obtained by using a crawl tool, like Screaming Frog. As with external links, the more internal links found referencing the redirect source, the more likely it is that the redirect should be kept. Unlike external links, internal links can almost always be updated so that the internal link does not reference the redirected source.

Table 8-4. Redirect Map—After Redirect Source Check

Redirect Source	Redirect Destination	...	Crawls	Entrances	External Links	Internal Links
/A	/B		50	0	56	6
/B	/C		64	0	42	0
/D	/E		0	0	0	0
/F	/B		2	4	3	158
/H	/Q		0	27	17	0

Bringing all these metrics together, as shown in Table 8-4, decisions can be made about what action to take for each redirect. The action will either be to keep the redirect as it is, keep but update the redirect destination, or remove the redirect.

- **Redirect /A -> /B**: This redirect destination needs to be updated per Step 2, but the question in this step is if the redirect is worth keeping based on the redirect source's activity. Robots are regularly crawling the redirect source URL, there are many external websites linking to that URL, and there are six internal links referencing the source URL. Given this data, the redirect source is still an important URL and should be kept, provided the updates noted in Step 2 are made. As well, those six internal links should be updated to reference the redirect destination, /C, instead.

- **Redirect /B -> /C**: This redirect is also worth keeping because it is regularly crawled, and there are several external links referencing the redirect source URL.

- **Redirect /D -> /E**: With no crawls, no entrances, and no links, this redirect should be removed from the website.

- **Redirect /F -> /B**: This redirect was noted for updates in Step 2, but after reviewing the data in this step, the redirect does not appear to be worth keeping as there are only two crawls, four entrances, and three external links. The only reason to keep the redirect would be if those three external links are highly valuable. Before removing this redirect, note there are several internal links referencing /F, and those should be updated to link to /C instead.

- **Redirect /H -> /Q**: While there are no crawls, there are 27 entrances tracked in the analytics tools. There are also a larger number of external links. Together, this likely means this redirect should be kept (and updated to a server-side redirect method if possible).

Step 4: Review Other Redirect Use Cases

After reviewing redirect destinations and sources, a final review should be made to see if there are any non-SEO reasons to keep the redirects. While a redirect source URL may not have any SEO-related activity, it might be used in an ad platform, or the redirect source URL might be printed in offline advertising. There could also be redirect sources that represent legacy URLs that would be worth maintaining on the off chance an old customer happened to use that legacy URL to return to the website. In the example table shown in Step 3, this might be the case for the redirect from /D and may justify keeping the redirect even though the SEO-related data indicated otherwise.

Regularly Checking Redirects

Outdated and inaccurate redirects are a form of technical debt. The more problems that exist within a website's redirects, the harder it becomes to update and maintain those redirects. This can make a redirect review and cleanup project a daunting task. Annual, quarterly, or monthly reviews of the website's redirects—depending on how active the website is—can help make this task more manageable and limit the technical debt. At a minimum, redirects should be fully reviewed and updated as part of any major update projects, like redesigns or redevelopments, to help improve the website's overall performance.

Measuring and Monitoring Guidelines: Redirects

- If a URL changes, what is the most appropriate place to redirect that URL to? If there is no relevant page to redirect the URL to, the URL should return an error message instead.

- Once redirects are added, monitor search results to see how quickly rankings are updated to use the new URL. As well, monitor ranking position for the new URL. Is the new URL ranking for the same terms and in the same position as the old URL or were rankings lost as a result of the redirect?

- Do any redirects on the website rely on client-side redirect methods? If so, are robots respecting the client-side redirects? If not, can the redirect be updated to a server-side method instead?

- Crawl the website to identify any internal links referencing redirected URLs, including links listed on the XML sitemap. Update the internal links to directly reference the final redirect destination. This helps improve crawl efficiency for robots and avoid robots indexing or ranking incorrect URLs.

- Crawl the website to identify any canonical tags that reference redirected URLs. Canonical tags should be updated to clearly establish which URLs should be indexed and rank in search results.

Conclusion: Tech SEO Audit

One of the most common ways of engaging with technical SEO is through an audit. An audit of the technical factors discussed in this book can, if done correctly, offer valuable insight into the various problems and opportunities present on a website. However, many audits end up being ineffective because the audit's focus is too formulaic—the audit becomes more about completing a checklist or reviewing key metrics instead of providing valuable insight.

Technical SEO is less about a checklist of items or certain metrics to review. Instead, it is about knowing which questions to ask as the website is reviewed. By asking those questions, the various problems or opportunities related to technical SEO will be uncovered. It is better to think of a technical SEO audit, then, as an investigation of the website.

An audit checklist, to the extent it is used at all, should only provide loose guidance for that investigation, but the investigation should be allowed to deviate depending on what types of issues are uncovered. With that in mind, and as a way to recap the issues discussed in this book, here are a few guiding questions to begin considering when auditing a website.

© Matthew Edgar 2023
M. Edgar, *Tech SEO Guide*, https://doi.org/10.1007/978-1-4842-9054-5_9

Question #1: Is the Website Crawlable?

At the most foundational level, robots need to be able to crawl the website and fetch content contained on the website's pages. If robots are unable to find all the pages on a website or are unable to load the pages once found, then a website will be unable to rank in search results.

- Are robots able to crawl the website or are there blocks preventing crawling? If blocks are in place, are those blocks created intentionally or accidentally?

- Can robots discover all the pages, images, and other files contained on the website? Are all files referenced in internal links and on the XML sitemap?

- Do link qualifiers (nofollow, sponsored, ugc) need to be used on the website to explain links? If so, are those qualifiers used appropriately?

- Are robots crawling all the files that should be crawled on the website? Which files, if any, are being missed?

- Can robots accurately render each page on the website? If not, what is preventing the page from being correctly rendered?

- Are URLs on the website well formed and used consistently? Is the canonical domain clearly defined and used consistently?

Question #2: Is the Website Indexable?

After crawling the website, robots analyze and assess the content found during the crawl, determining how to organize and index the content. Any quality issues within the content or errors found on the website can create problems with how the content is indexed, preventing pages from ranking.

- Are there not-found or server errors present on the website? Are those errors preventing robots from accurately indexing pages on the website?

- Which pages or other files are not being included in the search engine's index?

- Are there noindex directives used on the website? If so, are those directives used correctly and are those directives respected by search engines?

- What content-related issues are present on the website? What problems are those content-related issues creating for the website?

- Are canonical tags present? Are canonical tags used appropriately?

- Is the website presented in multiple languages or targeting multiple countries? If so, are hreflang tags used correctly? Are multiple languages listed correctly in the XML sitemap?

Question #3: Is the Website Ranking and How Is It Ranking?

Finally, search engine robots allow some pages and files to appear in search results. When a page appears in search results, it competes for attention with all other pages listed to earn a click from the person who conducted the search. Any problems or missed opportunities can result in lost clicks.

- Do all pages on the website use unique title tags? Are title tags correctly optimized?

- Are all pages that should rank in search results ranking? If pages are not ranking, are there technical reasons preventing pages from being able to rank (such as errors or misconfigured robot directives)?

- Are all pages ranking at the correct URLs? Do all ranking pages use the canonical domain or are some pages ranking using alternative domains (such as http instead of https)?

- Does the website comply with mobile-friendliness guidelines?

- Does the website load quickly and comply with Core Web Vitals?

- Is schema being used to mark up the content? If so, is it being displayed in Google's search results?

An Ongoing Process

A technical SEO audit will undoubtedly uncover many problems on a website. Not all of these issues can be or should be addressed at once. Remember, technical SEO is an ongoing process of improving the website's code and server configuration to better communicate with search engine robots.

Improving communication with robots is not instantaneous. This is a process that happens over weeks, months, and even years and includes the following:

- Addressing crawling issues, including making the website load faster, communicates to robots they can fetch more content from the website.

- Adding internal links or fixing issues in XML sitemap communicates to robots there are more pages to crawl and index on the website.

- Stating and enforcing canonical URL choices, including the canonical domain, sends a clear signal to robots about which pages to index.

- Using well-organized HTML and simplifying how pages are rendered improves how robots process and analyze a website's pages.

- Utilizing schema markup to better explain what information is available on the page and to allow search engines to present that content differently in search results.

- Alleviating robot confusion points by fixing not-found errors and redirect-related issues so that robots can make better choices about crawling, indexing, and ranking the website's pages.

By steadily addressing these and other issues, the communication with search engine robots will improve and so will performance in organic search results.

HTTP Response Status Codes

Whenever a file contained on a website is requested from the server, the server returns a numerical code that indicates the page's status, called an HTTP response status code. The status code indicates if the page is operating correctly, is in error, requires authentication, and more. The HTTP response status code is returned in the HTTP response header, which contains other details about the page. The first digit in the status code represents the class, while the other two digits provide more details. There are only five valid classes, 1 through 5. The HTTP status codes that have the greatest impact on SEO are listed in Table A-1.

© Matthew Edgar 2023
M. Edgar, *Tech SEO Guide*, https://doi.org/10.1007/978-1-4842-9054-5

Table A-1. HTTP Response Status Codes Most Useful for SEO

Class	Status Code	Reason Phrase	Notes
2	200	OK	The requested file has successfully loaded. This is what normal pages should return.
3	301	Moved Permanently	The requested file has permanently moved to a new location. Must include location in the header listing the URL to redirect to.
3	302	Found	The requested file has moved to a new location. Used for temporary redirects. Must list the URL to redirect to in the header's location field.
3	304	Not Modified	The requested file has not been modified since the previous request and the cached copy of the requested file can be used.
3	307	Temporary Redirect	Available in HTTP 1.1. The requested file has been temporarily redirected to a new location. Must list the URL to redirect to in the header's location field.
3	308	Permanent Redirect	Available in HTTP 1.1. The requested file has been permanently redirected to a new location. Must list the URL to redirect to in the header's location field.
4	401	Unauthorized	Authentication failed to access the requested file. Must include WWW-Authenticate in the header with authentication methods (or challenges).
4	403	Forbidden	The requestor has no permission to access the requested file.
4	404	Not Found	The requested file could not be located on the server.
4	410	Gone	The requested file has been purposefully removed from the server.
5	500	Internal Server Error	A generic error message indicating a server error occurred. Where possible, more specific status codes in the 5xx class should be used.
5	503	Service Unavailable	The server is currently unavailable. This is typically temporary.

References

Chapter 1: Crawling and Indexing

Google. "In-Depth Guide to How Google Search Works." Google Developers, October 13, 2022. https://developers.google.com/search/docs/advanced/guidelines/how-search-works.

Google. "Large site owner's guide to managing your crawl budget." Google Developers, October 13, 2022. https://developers.google.com/search/docs/advanced/crawling/large-site-managing-crawl-budget.

Google. "Qualify your outbound links to Google." Google Developers, October 11, 2022. https://developers.google.com/search/docs/advanced/guidelines/qualify-outbound-links.

Lahey, Connor. "Robots Meta Tag and X-Robots-Tag Explained." Semrush, December 16, 2020. www.semrush.com/blog/robots-meta/.

Moz. "Robots Meta Directives." Moz SEO Learning Center, June 2, 2022. https://moz.com/learn/seo/robots-meta-directives.

Moz. "Robots.txt." Moz SEO Learning Center, October 5, 2022. https://moz.com/learn/seo/robotstxt.

Weinstein, Mindy. "Where We Are Today With Google's Mobile-First Index." Search Engine Journal, July 9, 2022. www.searchenginejournal.com/google-mobile-first-indexing/346170/.

Chapter 2: URL and Domain Structure

MDN. "What is a URL?." October 8, 2022. https://developer.mozilla.org/en-US/docs/Learn/Common_questions/What_is_a_URL.

Montti, Roger. "Subdomains vs. Subfolders." Search Engine Journal, February 12, 2021. www.searchenginejournal.com/subdomains-vs-subfolders-seo/239795/.

Starr, Jeff. ".htaccess redirect to https and www." .htaccess made easy, March 22, 2016. https://htaccessbook.com/htaccess-redirect-https-www/.

SSL.com. "Redirect HTTP to HTTPS with Windows IIS 10." November 30, 2020. www.ssl.com/how-to/redirect-http-to-https-with-windows-iis-10/.

Chapter 3: Content Structure

Buckingham-Bullock, Alex. "Duplicate Content: SEO Best Practices to Avoid It." Semrush, December 23, 2020. www.semrush.com/blog/duplicate-content/.

Costello, Rachel. "Thin Pages." Lumar, October 30, 2018. www.deepcrawl.com/knowledge/technical-seo-library/thin-pages/.

Google. "Implement dynamic rendering." Google Developers, September 20, 2022. https://developers.google.com/search/docs/advanced/javascript/dynamic-rendering.

Google. "Infinite scroll search-friendly recommendations." Google Developers, February 13, 2014. https://developers.google.com/search/blog/2014/02/infinite-scroll-search-friendly.

Google. "Tell Google about localized versions of your page." Google Developers, September 16, 2022. https://developers.google.com/search/docs/advanced/crawling/localized-versions.

MDN. "Populating the page: how browsers work." October 6, 2022. https://developer.mozilla.org/en-US/docs/Web/Performance/How_browsers_work#render.

Moz. "Alt Text." Moz SEO Learning Center, August 31, 2022. https://moz.com/learn/seo/alt-text.

W3C. "Headings." Web Accessibility Initiative, May 4, 2017. www.w3.org/WAI/tutorials/page-structure/headings/.

Chapter 4: Schema and Structured Data Markup

Google. "Review snippet (Review, AggregateRating) structured data." Google Developers, October 13, 2022. https://developers.google.com/search/docs/appearance/structured-data/review-snippet.

IONOS. "Tag your website with RDFa according to Schema.org's guidelines." April 7, 2016. www.ionos.com/digitalguide/websites/website-creation/tutorial-rdfa-markup-with-schemaorg/.

MDN. "itemtype." September 14, 2022. https://developer.mozilla.org/en-US/docs/Web/HTML/Global_attributes/itemtype.

Schema.org. "Organization." March 17, 2022. https://schema.org/Organization.

Screaming Frog. "Structured Data Testing Using The SEO Spider Tool." June 7, 2022. www.screamingfrog.co.uk/structured-data-testing-validation/.

Chapter 5: Sitemaps

Bing. "Sitemaps." Bing Webmaster Tools. Accessed October 24, 2022. www.bing.com/webmasters/help/Sitemaps-3b5cf6ed.

Google. "Build and submit a sitemap." Google Developers, September 16, 2022. https://developers.google.com/search/docs/crawling-indexing/sitemaps/build-sitemap.

Google. "Tell Google about localized versions of your page." Google Developers, September 16, 2022. https://developers.google.com/search/docs/advanced/crawling/localized-versions.

Schwartz, Eli. "7 Reasons Why An HTML Sitemap Is Still A Must-Have." Search Engine Journal, November 30, 2021. www.searchenginejournal.com/html-sitemap-importance/325405/.

Chapter 6: Page Experience: Core Web Vitals and More

alphadigital. "Responsive vs Adaptive vs Dedicated Mobile Sites: Which Should You Use?". Alpha Digital, February 2, 2018. www.alphadigital.com.au/blog/advice/responsive-adaptive-dedicated-mobile-sites/.

Google. "About PageSpeed Insights." Google Developers, May 10, 2022. https://developers.google.com/speed/docs/insights/v5/about.

Google. "Get started with mobile-friendliness." Google Developers, April 21, 2022. https://developers.google.com/search/mobile-sites/get-started.

Jessier, Myriam. "Intrusive Interstitials: Guidelines To Avoiding Google's Penalty." Smashing Magazine, May 9, 2017. www.smashingmagazine.com/2017/05/intrusive-interstitials-guidelines-avoid-google-penalty/.

web.dev. "Document doesn't use legible font sizes." August 20, 2019. `https://web.dev/font-size/`.

web.dev. "Tap targets are not sized appropriately." May 2, 2019. `https://web.dev/tap-targets/`.

Chapter 7: Not-Found Errors

Justinmind. "6 best practices for 404 pages with killer UX." UXPlanet, April 13, 2017. `https://uxplanet.org/6-best-practices-for-404-pages-with-killer-ux-d9305db19ad9`.

Nesterets, Julia. "How to find links to broken pages." JetOctopus, June 15, 2020. `https://jetoctopus.com/how-to-find-links-to-broken-pages/`.

Sharp, Dan. "How To Find Broken Links Using the SEO Spider." Screaming Frog, October 22, 2021. `www.screamingfrog.co.uk/broken-link-checker/`.

Sissons, Oliver. "404 vs 410 - The Technical SEO Experiment." Reboot, September 12, 2022. `www.rebootonline.com/blog/404-vs-410-the-technical-seo-experiment/`.

Chapter 8: Redirects

Hardwick, Joshua. "301 vs. 302 Redirects for SEO: Which Should You Use?." Ahrefs, August 3, 2022. `https://ahrefs.com/blog/301-vs-302-redirects/`.

Google. "Redirects and Google Search." Google Developers, September 22, 2022. `https://developers.google.com/search/docs/crawling-indexing/301-redirects`.

W3Schools. "How To Redirect to Another Webpage." Accessed October 24, 2022. `www.w3schools.com/howto/howto_js_redirect_webpage.asp`.

Appendix A: HTTP Response Status Codes

Fielding, Roy T., Mark Nottingham, and Julian Reschke. "RFC 9110." RFC 9110: HTTP Semantics, June 2022. `www.rfc-editor.org/rfc/rfc9110.html#name-status-codes`.

I

Index

© Matthew Edgar 2023

M. Edgar, *Tech SEO Guide*, https://doi.org/10.1007/978-1-4842-9054-5

X, Y, Z